Selected Papers from 43rd International Conference of Theoretical Physics

Selected Papers from 43rd International Conference of Theoretical Physics

Matter to the Deepest, Recent Developments In Physics Of Fundamental Interactions (MTTD2019)

Special Issue Editors

Janusz Gluza
Jan Sładkowski
Bartosz Dziewit
Ilona Bednarek

MDPI • Basel • Beijing • Wuhan • Barcelona • Belgrade • Manchester • Tokyo • Cluj • Tianjin

Special Issue Editors

Janusz Gluza
University of Silesia in Katowice
Poland

Jan Sładkowski
University of Silesia in Katowice
Poland

Bartosz Dziewit
University of Silesia in Katowice
Poland

Ilona Bednarek
University of Silesia in Katowice
Poland

Editorial Office
MDPI
St. Alban-Anlage 66
4052 Basel, Switzerland

This is a reprint of articles from the Special Issue published online in the open access journal *Symmetry* (ISSN 2073-8994) (available at: https://www.mdpi.com/journal/symmetry/special_issues/mttd2019_symmetry).

For citation purposes, cite each article independently as indicated on the article page online and as indicated below:

LastName, A.A.; LastName, B.B.; LastName, C.C. Article Title. *Journal Name* **Year**, *Article Number*, Page Range.

ISBN 978-3-03928-370-5 (Pbk)
ISBN 978-3-03928-371-2 (PDF)

© 2020 by the authors. Articles in this book are Open Access and distributed under the Creative Commons Attribution (CC BY) license, which allows users to download, copy and build upon published articles, as long as the author and publisher are properly credited, which ensures maximum dissemination and a wider impact of our publications.

The book as a whole is distributed by MDPI under the terms and conditions of the Creative Commons license CC BY-NC-ND.

Contents

About the Special Issue Editors vii

Preface to "Selected Papers from 43rd International Conference of Theoretical Physics" ix

Bartosz Dziewit, Jacek Holeczek, Sebastian Zając and Marek Zrałek
Family Symmetries and Multi Higgs Doublet Models
Reprinted from: *Symmetry* **2020**, *12*, 156, doi:10.3390/sym12010156 1

Janusz Gluza, Magdalena Kordiaczyńska and Tripurari Srivastava
Doubly Charged Higgs Bosons and Spontaneous Symmetry Breaking at eV and TeV Scales
Reprinted from: *Symmetry* **2020**, *12*, 153, doi:10.3390/sym12010153 9

Zoltán Trócsányi
Super-Weak Force and Neutrino Masses
Reprinted from: *Symmetry* **2020**, *12*, 107, doi:10.3390/sym12010107 25

Torsten Asselmeyer-Maluga, Jerzy Król and Tomasz Miller
Does Our Universe Prefer Exotic Smoothness?
Reprinted from: *Symmetry* **2020**, *12*, 98, doi:10.3390/sym12010098 43

Hong Zhang
Axion Stars
Reprinted from: *Symmetry* **2020**, *12*, 25, doi:10.3390/sym12010025 55

Simonas Draukšas, Vytautas Dudenas, Thomas Gajdosik, Andrius Juodagalvis, Paulius Juodsnukis and Darius Jurčiukonis
The Grimus–Neufeld Model with FlexibleSUSY at One-Loop
Reprinted from: *Symmetry* **2019**, *11*, 1418, doi:10.3390/sym11111418 65

Stanisław Jadach, Wiesław Płaczek and Maciej Skrzypek
Exponentiation in QED and Quasi-Stable Charged Particles
Reprinted from: *Symmetry* **2019**, *11*, 1389, doi:10.3390/sym11111389 73

About the Special Issue Editors

Janusz Gluza works at the Institute of Physics of the Faculty of Science and Technology of the University of Silesia in Katowice. He deals with particle physics and their interactions. Scholarship holder of the Foundation for Polish Science and the Alexander von Humboldt Foundation, a three-year postdoc at the DESY-Zeuthen (Germany). He collaborates with scientists from leading scientific centers in the world (Germany, Switzerland, France, Italy, Spain, USA) and in Poland (Institute of Nuclear Physics of the Polish Academy of Sciences in Krakow). For several years involved in cooperation with CERN in Geneva and the Future Circular Collider collaboration. An active member of international scientific networks, principal investigator of several grants of the National Science Center in Poland.

Jan Sladkowski was born in Swietochłowice, Poland. He earned his Ph.D. and habilitation in theoretical physics from the University of Silesia, He is notable for his work on the role of exotic smoothness in cosmology, quantum game theory, and econophysics. He has held visiting posts at the Bielefeld University (A. von Humboldt Fellow) and at the University of Wisconsin–Madison. He presently holds a professor position at the University of Silesia. He has published about a hundred papers on quantum field theory, mathematical physics, quantum information processing, and econophysics.

Bartosz Dziewit was born in Sosnowiec in 1980. In 2014, he received his Ph.D. degree in Physics from the University of Silesia in Katowice. He is an assistant professor in the Faculty of Science and Technology at his alma mater. His research interests are mainly focused on the theoretical aspects of particle physics, with specific emphasis on neutrino physics, in addition to computer sciences and data analysis.

Ilona Bednarek was born in Ruda Slaska, Poland. She obtained her Ph.D. and habilitation in physics from the University of Silesia. Her main scientific interests are related to the physics and astrophysics of neutron stars. Her research problems in particular, concern theoretical models of the matter of neutron stars with non-zero strangeness. She has published over 50 scientific papers in this field of research.

Preface to "Selected Papers from 43rd International Conference of Theoretical Physics"

This Special Issue consists of 7 papers based on talks given at the 43rd International Conference of Theoretical Physics: Matter to the Deepest, Recent Developments in Physics Of Fundamental Interactions. The conference took place in Chorzów in an Upper Silesia region of Poland from the 1st to the 6th of September 2019.

Conference web page with Indico materials from all talks can be found at the following link: http://indico.if.us.edu.pl/event/5/overview.

We thank the speakers for their effort in preparing talks interesting to all audiences. Especially we are grateful to those who found time to prepare their contributions for publishing in this Special Issue.

We highly appreciate the efficient collaboration with the *Symmetry* Editorial team in producing this Special Issue of proceedings.

Janusz Gluza, Jan Sładkowski, Bartosz Dziewit, Ilona Bednarek
Special Issue Editors

Article

Family Symmetries and Multi Higgs Doublet Models

Bartosz Dziewit [1,*], Jacek Holeczek [1], Sebastian Zając [2] and Marek Zrałek [3]

[1] August Chełkowski Institute of Physics, University of Silesia in Katowice, ul. 75 Pułku Piechoty 1, 41-500 Chorzów, Poland
[2] SGH Warsaw School of Economics, Collegium of Economic Analysis, ul. Madalińskiego 6/8, 02-513 Warsaw, Poland
[3] Faculty of Applied Sciences, Humanitas University, ul. Kilińskiego 43, 41-200 Sosnowiec, Poland
* Correspondence: Bartosz.Dziewit@us.edu.pl

Received: 30 November 2019; Accepted: 9 January 2020; Published: 12 January 2020

Abstract: Imposing a family symmetry on the Standard Model in order to reduce the number of its free parameters, due to the Schur's Lemma, requires an explicit breaking of this symmetry. To avoid the need for this symmetry to break, additional Higgs doublets can be introduced. In such an extension of the Standard Model, we investigate family symmetries of the Yukawa Lagrangian. We find that adding a second Higgs doublet (2HDM) does not help, at least for finite subgroups of the $U(3)$ group up to the order of 1025.

Keywords: lepton masses and mixing; family symmetry

1. Introduction

At currently achievable energies, the Standard Model (SM) of fundamental interactions is a very good working theory. However, it is commonly believed that it is only an effective theory which at higher energies need to be modified. One of the signs of this state of things is a large number of free parameters (more than 20) which now need to be fitted from experiments. The main parameters are: masses, mixing angles and CP violating phases for quarks and leptons. The SM does not explain those parameters but introduces the mechanism by means of which all particles acquire masses by the so called Higgs mechanism. One of several proposals how to restrict number of a free parameters in the SM is to introduce symmetry between Yukawa constants in Yukawa SM interaction in such a way that after spontaneous symmetry breaking get masses and mixing matrix parameters for quarks and leptons which are consistent with experience. Such symmetry is known in the literature as a flavor symmetry [1] (but also family or horizontal symmetry). Lepton sector and especially neutrino physics is an attractive area to search for such a symmetry due to the so called lepton mixing matrix [2,3].

There exist direct links between the mixing and lepton masses. Charged lepton and neutrino mass matrices $M_{l(\nu)}$ are diagonalized by biunitary transformations (for Majorana neutrinos $(U_\nu)_R = (U_\nu)_L^*$):

$$(U_{l(\nu)})_L^\dagger M_{l(\nu)} (U_{l(\nu)})_R . \tag{1}$$

The lepton mixing matrix U_{PMNS} is composed from the charged lepton $(U_l)_L$ and neutrino $(U_\nu)_L$ diagonalizing matrices:

$$U_{PMNS} = (U_l)_L^\dagger (U_\nu)_L . \tag{2}$$

Elements of matrix (2) are determined in various of neutrino experiments. When we impose a family symmetry in the ordinary not extended SM, we obtain [4,5]:

$$A_L^{i\dagger}(M_l M_l^\dagger) A_L^i = (M_l M_l^\dagger), \tag{3}$$

$$A_L^{i\dagger}(M_\nu M_\nu^\dagger) A_L^i = (M_\nu M_\nu^\dagger), \tag{4}$$

where:

$$A_L^i = A_L(g_i), \quad i = 1, 2, \ldots, N, \tag{5}$$

are 3 dimensional representations matrices for the left handed lepton doublets for some N-order flavour symmetry group $\mathcal{G}$.

As a direct consequence of the Schur's Lemma—since $M_l M_l^\dagger$ and $M_\nu M_\nu^\dagger$ are proportional to the identity matrix, the lepton mixing matrix U_{PMNS} becomes trivial.

In the literature there are some ideas about how to escape from the trivialisation of a matrix (2). One approach is to break the family symmetry group by scalar singlet—so called "flavons" (e.g., [6,7]). Non trivial mixing can be also achieved by extending the Higgs sector by additional multiplets (e.g., [8,9]). A proposal for the two Higgs doublet model (2HDM) [10] was widely discussed in [11]. In the most general situation, the mass generation mechanism allows couplings with various Higgs fields. In this context, extensions of the SM assuming the existence of different numbers of doublets and Higgs triplets are allowed. Theoretical proposals assuming the existence of two Higgs fields are not the only possible and potentially experimentally verifiable space for applying the symmetry implementation. In this paper, the methodology proposed previously for 2HDM only [11] is extended to any number of additional Higgs fields.

Additionally, new forms of results, equivalent to [11], for 2HDM are given. The obtained results depend on many phases and we present here a more detailed discussion concerning relations between them. We hope that it may help to determine the analytical origin of these solutions.

2. Multi Higgs Doublet Description

Discussion below stands for Dirac neutrinos. To describe the coupling between lepton fields and the Higgs field we take the n-Higgs doublet Yukawa interaction term of the form:

$$L_Y = -(h_i^l)_{\alpha\beta} \bar{L}_{\alpha L} \Phi_i l_{\beta R} - (h_i^\nu)_{\alpha\beta} \bar{L}_{\alpha L} \tilde{\Phi}_i \nu_{\beta R} + \text{H.c.}, \tag{6}$$

where $i = 1, 2, \ldots, n$ and $\alpha, \beta = e, \mu, \tau$.

The charged lepton states $l_{\beta R}$ and neutrinos $\nu_{\beta R}$ are right-handed $SU(2)$ singlets and then the gauge doublets for the left-handed lepton and Higgs fields are:

$$L_{\alpha L} = \begin{pmatrix} \nu_{\alpha L} \\ l_{\alpha L} \end{pmatrix}, \tag{7}$$

$$\Phi_i = \begin{pmatrix} \varphi_i^0 \\ \varphi_i^- \end{pmatrix}, \quad \tilde{\Phi}_i = \begin{pmatrix} \varphi_i^{-*} \\ -\varphi_i^{0*} \end{pmatrix} = i\sigma_2 \Phi_i^*, \tag{8}$$

where φ_i^0 and φ_i^- are complex scalar fields in spacetime for $i = 1, 2, \ldots, n$ and $\sigma_2 = \begin{pmatrix} 0 & -i \\ i & 0 \end{pmatrix}$.

The 3×3 Yukawa matrices h_i^l and h_i^ν each define the couplings of left-handed doublets with right-handed singlets via the i-th Higgs doublet. Due to the form of the Higgs potentials, ground states occur at non-zero φ, with the vacuum expectation values:

$$<\Phi_i> = \frac{1}{\sqrt{2}} \begin{pmatrix} v_i \\ 0 \end{pmatrix} \quad \text{and} \quad <\tilde{\Phi}_i> = \frac{1}{\sqrt{2}} \begin{pmatrix} -v_i^* \\ 0 \end{pmatrix}, \tag{9}$$

for some complex-valued v_i, where:

$$\sqrt{|v_1|^2 + |v_2|^2 + \cdots + |v_n|^2} = (\sqrt{2}G_F)^{-1/2} \sim 246 \text{ GeV}. \tag{10}$$

Mass matrices for charged leptons and neutrinos read as follows:

$$M_l = -\frac{1}{\sqrt{2}}(v_1^* h_1^l + \cdots + v_n^* h_n^l), \tag{11}$$

$$M_\nu = \frac{1}{\sqrt{2}}(v_1 h_1^\nu + \cdots + v_n h_n^\nu). \tag{12}$$

For some finite flavour group $\mathcal{G}$, the family symmetry means that after fields transformations (A_L, A_l^R and A_ν^R are 3 dimensional representations):

$$L_{\alpha L} \to L'_{\alpha L} = (A_L)_{\alpha,\chi} L_{\chi L}, \tag{13}$$

$$l_{\beta R} \to l'_{\beta R} = \left(A_l^R\right)_{\beta,\delta} l_{\delta R}, \tag{14}$$

$$\nu_{\beta R} \to \nu'_{\beta R} = \left(A_\nu^R\right)_{\beta,\delta} \nu_{\delta R}, \tag{15}$$

and (A_Φ is a n dimensional representation):

$$\Phi_i \to \Phi'_i = (A_\Phi)_{ik} \Phi_k, \tag{16}$$

the Lagrangian does not change:

$$\mathcal{L}(L_{\alpha L}, l_{\beta R}, \nu_{\beta R}, \Phi_i) = \mathcal{L}(L'_{\alpha L}, l'_{\beta R}, \nu'_{\beta R}, \Phi'_i). \tag{17}$$

Symmetry conditions can be written as an eigenproblem of a direct product of unitary group representations to the eigenvalue 1. For any group elements we have:

$$\left((A_\Phi)^\dagger \otimes (A_L)^\dagger \otimes (A_l^R)^T\right)_{k,\alpha,\delta;i,\beta,\gamma} (h_i^l)_{\beta,\gamma} = (h_k^l)_{\alpha,\delta}, \tag{18}$$

$$\left((A_\Phi)^T \otimes (A_L)^\dagger \otimes (A_\nu^R)^T\right)_{k,\alpha,\delta;i,\beta,\gamma} (h_i^\nu)_{\beta,\gamma} = (h_k^\nu)_{\alpha,\delta}. \tag{19}$$

It is sufficient to check the above equations for group generators only as then they will automatically be satisfied for all group elements.

In such a model, the invariance equations for the mass matrices are not trivial, so we avoid the consequences of Schur's Lemma:

$$A_L M^{l(\nu)} \left(A_{l(\nu)}^R\right)^\dagger = \frac{1}{\sqrt{2}} \sum_{i,k=1}^n h_i^{l(\nu)} (A_\Phi)_{i,k} v_k \neq M^{l(\nu)}. \tag{20}$$

The same conclusion is valid if one assumes that neutrinos have Majorana nature. In such a frame, the Yukawa interaction Lagrangian has to be rewritten in an appropriate way (see [11]) producing family symmetry condition in the form:

$$\left((A_\Phi)^T \otimes (A_\Phi)^T \otimes (A^L)^\dagger \otimes (A^L)^\dagger\right)_{k,m,\chi,\eta,i,j,\alpha,\beta} (h_{ij}^\nu)_{\alpha,\beta} = (h_{km}^\nu)_{\chi,\eta}. \tag{21}$$

3. Two Higgs Doublet Model (2HDM) Results

As a potential flavour symmetry group $\mathcal{G}$, we chose finite, non-abelian subgroups of $U(3)$, up to the order of 1025. This class of groups is very important in practice [12], even though there exist models in which the flavour symmetry group cannot be conceived as a subgroup of $U(3)$ (the upper limit on the group order was of course due to the calculation time). Using the GAP [13] system for computational discrete algebra, with the included SMALL GROUPS LIBRARY [14] and the REPSN [15] package for constructing representations of finite groups, we have found groups which fulfil the requirements of our model and impose flavour symmetry on the Yukawa Lagrangian. Next we have calculated the Yukawa matrices and created mass matrices and mixing matrices. The last step was to check agreement with experimental data. In total we have found 10862 groups with at least one 2 dimensional and at least one 3 dimensional irreducible representation. Only 413 of these groups are subgroups of $U(3)$. Either a group has at least one faithful 3 dimensional irreducible representation (there are 173 such groups) or it has at least one faithful 1+2 reducible representation (there are 240 such groups).

3.1. Results for Dirac Combinations

All obtained solutions for Yukawa matrices for Dirac neutrinos and charged leptons can be expressed through seven base forms. Putting $\omega = e^{2\pi i k/3}$ and allowing any integer k value (note that $\omega^3 = 1$ and $\omega^2 = \omega^*$), the first three forms are:

$$h_1 = \begin{pmatrix} 0 & 0 & 1 \\ \omega^2 & 0 & 0 \\ 0 & \omega & 0 \end{pmatrix}, \quad h_2 = \begin{pmatrix} 0 & 1 & 0 \\ 0 & 0 & \omega \\ \omega^2 & 0 & 0 \end{pmatrix}. \tag{22}$$

The next three forms can be obtained from the above ones through a simple interchange $h_1 \rightleftarrows h_2$. The last, seventh form, valid only for $k = \ldots, -2, 1, 4, \ldots$, is (note the diagonality of these matrices):

$$h_1 = \begin{pmatrix} 1 & 0 & 0 \\ 0 & \omega^2 & 0 \\ 0 & 0 & \omega \end{pmatrix}, \quad h_2 = h_1^*. \tag{23}$$

Dirac neutrinos are always defined by "ordered" pairs:

$$\{h_1^\nu, h_2^\nu\} = \{h_1, e^{i\phi} h_2\}, \tag{24}$$

where ϕ are some real phases (which depend on the actual group and its representations' combinations). For each of such Dirac neutrinos' solutions, there always exist two different solutions for charged leptons, defined by the two corresponding "ordered" pairs:

$$\{h_1^l, h_2^l\} = \{h_2, e^{-i(\delta_l + \phi)} h_1\}, \tag{25}$$

where $\delta_l = 0, \pi$.

Assuming complex c_x, real v_1, ϕ_1, v_2, ϕ_2 and putting:

$$M = c_x \left[v_1 e^{i\phi_1} h_1 + v_2 e^{i\phi_2} h_2 \right], \tag{26}$$

we get exactly the same set of three eigenvalues of $MM^\dagger$ for any of the seven above Yukawa matrices forms (note that this set of eigenvalues is invariant with respect to $v_1 \rightleftarrows v_2$ and/or $\phi_1 \rightleftarrows \phi_2$, hence we can safely use them for all cases $h_1 \rightleftarrows h_2$, and that if $v_1 v_2 \geq 0$ then $m_1^2 \leq m_2^2 \leq m_3^2$ when $\pi \leq \phi_2 - \phi_1 \leq 4\pi/3$ and $m_1^2 \geq m_2^2 \geq m_3^2$ when $0 \leq \phi_2 - \phi_1 \leq \pi/3$):

$$\text{eigenvalues}\,(M M^\dagger) = \begin{pmatrix} m_1^2 \\ m_2^2 \\ m_3^2 \end{pmatrix} = |c_x|^2 \begin{pmatrix} v_1^2 + v_2^2 + 2 v_1 v_2 \cos(\phi_2 - \phi_1) \\ v_1^2 + v_2^2 + 2 v_1 v_2 \cos(\phi_2 - \phi_1 - 2\pi/3) \\ v_1^2 + v_2^2 + 2 v_1 v_2 \cos(\phi_2 - \phi_1 + 2\pi/3) \end{pmatrix}. \quad (27)$$

The mass squared differences are (note that if $v_1 v_2 \geq 0$ then $\Delta m_{31}^2 \geq \Delta m_{32}^2 \geq \Delta m_{21}^2 \geq 0$ when $7\pi/6 \leq \phi_2 - \phi_1 \leq 4\pi/3$ and $\Delta m_{31}^2 \leq \Delta m_{32}^2 \leq \Delta m_{21}^2 \leq 0$ when $\pi/6 \leq \phi_2 - \phi_1 \leq \pi/3$):

$$\begin{aligned} \Delta m_{21}^2 &= +2\sqrt{3} |c_x|^2 v_1 v_2 \sin(\phi_2 - \phi_1 - \pi/3), \\ \Delta m_{31}^2 &= -2\sqrt{3} |c_x|^2 v_1 v_2 \sin(\phi_2 - \phi_1 + \pi/3), \\ \Delta m_{32}^2 &= -2\sqrt{3} |c_x|^2 v_1 v_2 \sin(\phi_2 - \phi_1). \end{aligned} \quad (28)$$

In all cases $M M^\dagger = M^\dagger M$ and then for the seventh form $M M^\dagger = \text{diag}\,(m_1^2, m_2^2, m_3^2)$ (so, no neutrino mixing is possible at all), while for the first six forms it is:

$$M M^\dagger = |c_x|^2 \begin{pmatrix} v_1^2 + v_2^2 & v_1 v_2 e^{-i(\phi_2 - \phi_1 + 2\pi k/3)} & v_1 v_2 e^{i(\phi_2 - \phi_1 - 2\pi k/3)} \\ v_1 v_2 e^{i(\phi_2 - \phi_1 + 2\pi k/3)} & v_1^2 + v_2^2 & v_1 v_2 e^{-i(\phi_2 - \phi_1)} \\ v_1 v_2 e^{-i(\phi_2 - \phi_1 - 2\pi k/3)} & v_1 v_2 e^{i(\phi_2 - \phi_1)} & v_1^2 + v_2^2 \end{pmatrix}, \quad (29)$$

and the unitary matrix that diagonalizes it (so that $U^\dagger M M^\dagger U = \text{diag}\,(m_1^2, m_2^2, m_3^2)$) is:

$$U = \frac{1}{\sqrt{3}} \begin{pmatrix} e^{-2\pi i k/3} & e^{-2\pi i (k-1)/3} & e^{-2\pi i (k+1)/3} \\ 1 & e^{-2\pi i/3} & e^{2\pi i/3} \\ 1 & 1 & 1 \end{pmatrix}. \quad (30)$$

Note that the U matrix does not depend on the phase difference $\phi_2 - \phi_1$ at all so, it will be exactly the same for neutrinos $(U_\nu)_L$ and charged leptons $(U_l)_L$. Hence, the $U_{PMNS} = (U_l)_L^\dagger (U_\nu)_L = I$ (so again, no neutrino mixing is possible at all).

In order to directly apply the above equations to Dirac neutrinos, one should put $c_x \to c_\nu$, $v_1 \to v_1$, $\phi_1 \to \phi_1$, $v_2 \to v_2$, $\phi_2 \to \phi + \phi_2$ (so $\phi_2 - \phi_1 \to \phi + \phi_2 - \phi_1$), while for charged leptons, one should put $c_x \to c_l$, $v_1 \to v_2$, $\phi_1 \to -(\delta_l + \phi + \phi_2)$, $v_2 \to v_1$, $\phi_2 \to -\phi_1$ (so $\phi_2 - \phi_1 \to \delta_l + \phi + \phi_2 - \phi_1$), where we assume that the vacuum expectation values are $v_1 e^{i \phi_1}$ and $v_2 e^{i \phi_2}$ (the same for neutrinos and charged leptons, of course). When moving between Dirac neutrinos and charged leptons, we can easily notice that all equations are invariant with respect to $v_1 \rightleftarrows v_2$ and the phase difference $\phi_2 - \phi_1$ is simply shifted by $\delta_l = 0, \pi$. That means that, for $\delta_l = \pi$, all mass squared differences Δm_{ij}^2 will change signs, so that their mass ordering schemes will be reversed, while for $\delta_l = 0$ there will be no change at all.

The ratios of experimental mass squared differences for neutrinos and charged leptons are:

$$\begin{aligned} |\Delta m_{atm}^2 / \Delta m_{sol}^2| &\approx (2.4\text{--}2.6) \times 10^{-3} / (7.4\text{--}7.7) \times 10^{-5} \approx 33 \pm 3, \\ (m_\tau^2 - m_e^2) / (m_\mu^2 - m_e^2) &\approx 282.8, \end{aligned} \quad (31)$$

and so they cannot be reproduced in this theoretical frame (they would need to be exactly equal while they differ by a factor of about 8 to 9).

Moreover, Equation (27) are not able to reproduce experimental masses of charged leptons at all. They return, in all cases, $1 \leq m_\tau/m_\mu \leq 2$ and requiring $m_\mu/m_e \approx 206.8$ or $m_\tau/m_e \approx 3477$ returns $m_\tau \approx m_\mu$.

3.2. Results for Majorana Combinations

All obtained solutions for Majorana neutrinos, which do not produce a scalar mass matrix M_ν, are defined by "ordered" quadruples:

$$\{h_{11}^\nu, h_{12}^\nu, h_{21}^\nu, h_{22}^\nu\} = \{h_2, v_0 e^{i(\phi+\phi_0)} I_3, v_0 e^{i(\delta+\phi+\phi_0)} I_3, e^{i(\delta+2\phi)} h_1\}, \quad (32)$$

where h_1 and h_2 are defined by Equation (23), ϕ are some real phases (which depend on the actual group and its representations' combinations), $\delta = 0, \pi$ and $v_0 e^{i\phi_0}$ is a free complex parameter. For each of such Majorana neutrinos' solutions, there always exist two different solutions for charged leptons, defined by the two corresponding "ordered" pairs $\{h_1^l, h_2^l\} = \{h_2, e^{-i(\delta_l+\phi)} h_1\}$, where $\delta_l = 0, \pi$ (note that δ_l is independent of δ so, there always exist four different combinations for each ϕ).

Assuming complex c_ν (usually denoted as $g/(2M)$), real $v_0, \phi_0, v_1, \phi_1, v_2, \phi_2$ and putting (where we assume that the vacuum expectation values are $v_1 e^{i\phi_1}$ and $v_2 e^{i\phi_2}$):

$$\begin{aligned} M_\nu &= c_\nu \left[v_1^2 e^{2i\phi_1} h_{11} + v_1 v_2 e^{i(\phi_1+\phi_2)} (h_{12}+h_{21}) + v_2^2 e^{2i\phi_2} h_{22} \right] \\ &= c_\nu \left[v_2^2 e^{i[\delta+2(\phi+\phi_2)]} h_1 + v_1^2 e^{2i\phi_1} h_2 + v_0 v_1 v_2 \left(1 + e^{i\delta}\right) e^{i(\phi+\phi_0+\phi_1+\phi_2)} I_3 \right], \end{aligned} \quad (33)$$

we find that the neutrino mass matrix M_ν is a diagonal matrix and $M_\nu M_\nu^\dagger = M_\nu^\dagger M_\nu = \text{diag}\left(m_1^2, m_2^2, m_3^2\right)$, where:

$$\begin{pmatrix} m_1^2 \\ m_2^2 \\ m_3^2 \end{pmatrix} = |c_\nu|^2 \begin{pmatrix} v_1^4 + v_2^4 + 2v_1^2 v_2^2 \cos[\delta + 2(\phi+\phi_2-\phi_1)] \\ + 4 v_0 v_1 v_2 \left[v_1^2 \cos(\phi+\phi_2-\phi_1+\phi_0) \right. \\ \left. + v_2^2 \cos(\phi+\phi_2-\phi_1-\phi_0) + v_0 v_1 v_2 \right] \cos(\delta/2) \\[6pt] v_1^4 + v_2^4 + 2v_1^2 v_2^2 \cos[\delta + 2(\phi+\phi_2-\phi_1) + 2\pi/3] \\ + 4 v_0 v_1 v_2 \left[v_1^2 \cos(\phi+\phi_2-\phi_1+\phi_0-2\pi/3) \right. \\ \left. + v_2^2 \cos(\phi+\phi_2-\phi_1-\phi_0-2\pi/3) + v_0 v_1 v_2 \right] \cos(\delta/2) \\[6pt] v_1^4 + v_2^4 + 2v_1^2 v_2^2 \cos[\delta + 2(\phi+\phi_2-\phi_1) - 2\pi/3] \\ + 4 v_0 v_1 v_2 \left[v_1^2 \cos(\phi+\phi_2-\phi_1+\phi_0+2\pi/3) \right. \\ \left. + v_2^2 \cos(\phi+\phi_2-\phi_1-\phi_0+2\pi/3) + v_0 v_1 v_2 \right] \cos(\delta/2) \end{pmatrix}. \quad (34)$$

The mass squared differences are:

$$\begin{aligned} \Delta m_{21}^2 &= 2\sqrt{3} |c_\nu|^2 v_1 v_2 \{ -v_1 v_2 \sin[\delta + 2(\phi+\phi_2-\phi_1) + \pi/3] \\ &\quad + 2 v_0 \left[v_1^2 \sin(\phi+\phi_2-\phi_1+\phi_0-\pi/3) + v_2^2 \sin(\phi+\phi_2-\phi_1-\phi_0-\pi/3)\right] \cos(\delta/2) \}, \\ \Delta m_{31}^2 &= 2\sqrt{3} |c_\nu|^2 v_1 v_2 \{ +v_1 v_2 \sin[\delta + 2(\phi+\phi_2-\phi_1) - \pi/3] \\ &\quad - 2 v_0 \left[v_1^2 \sin(\phi+\phi_2-\phi_1+\phi_0+\pi/3) + v_2^2 \sin(\phi+\phi_2-\phi_1-\phi_0+\pi/3)\right] \cos(\delta/2) \}, \\ \Delta m_{32}^2 &= 2\sqrt{3} |c_\nu|^2 v_1 v_2 \{ +v_1 v_2 \sin[\delta + 2(\phi+\phi_2-\phi_1)] \\ &\quad - 2 v_0 \left[v_1^2 \sin(\phi+\phi_2-\phi_1+\phi_0) + v_2^2 \sin(\phi+\phi_2-\phi_1-\phi_0)\right] \cos(\delta/2) \}. \end{aligned} \quad (35)$$

For $\delta = \pi$ ($e^{i\delta} = -1$), we can also reuse Equations (27) and (28), when we put $c_x \to c_\nu$, $v_1 \to v_2^2$, $\phi_1 \to \pi + 2(\phi+\phi_2)$, $v_2 \to v_1^2$, $\phi_2 \to 2\phi_1$ (so $\phi_2 - \phi_1 \to \pi - 2(\phi+\phi_2-\phi_1)$). Note that the mass ordering schemes will be reversed between $\delta = 0$ and $\delta = \pi$, which can easily be seen if one puts $v_0 = 0$ in Equations (34) and (35). For the corresponding charged leptons, we can also simply reuse Equations (27) and (28), where we put $c_x \to c_l$, $v_1 \to v_2$, $\phi_1 \to -(\delta_l + \phi + \phi_2)$, $v_2 \to v_1$, $\phi_2 \to -\phi_1$ (so $\phi_2 - \phi_1 \to \delta_l + \phi + \phi_2 - \phi_1$). As already mentioned, Equation (27) is not able to reproduce experimental masses of charged leptons at all. Moreover, as all relevant matrices are diagonal (for both neutrinos and charged leptons), no neutrino mixing is possible at all.

4. Conclusions

Multiple Higgs doublet models are in general promising that, in order to get a non trivial lepton mixing matrix, one will not need to explicitly break the family symmetry. However, our results for the 2HDM are utterly negative. The big open question is why. First, we select finite, non-abelian subgroups of $U(3)$ which are provided by the SMALL GROUPS LIBRARY [14] in GAP [13]. Is it possible that we miss some vital groups (due to the constrains that we impose when selecting groups or due to the used library itself)? Then, is it possible that non trivial solutions can only be obtained for models with more than two Higgs doublets? Finally, is it possible that the Equations (18), (19) and (21) will always lead to only trivial solutions, even though these models seem to be free from the consequences of the Schur's Lemma (20)? In order to address, at least partially, the last two questions, we are currently working on processing groups which are not subgroups of $U(3)$ for 2HDM and on a family symmetry approach with three Higgs doublets, hoping that it will give some positive outcome.

Author Contributions: All authors contributed equally to the reported research. All authors have read and agreed to the published version of the manuscript.

Funding: This work has been supported by the Polish National Science Centre (NCN) under Grant No. UMO—2013/09/B/ST2/03382.

Conflicts of Interest: Authors declare no conflict of interest.

Abbreviations

The following abbreviations are used in this manuscript:

SM   Standard Model
PMNS   Pontecorvo–Maki–Nakagawa–Sakata
2HDM   Two Higgs doublet model

References

1. King, S.F.; Luhn, C. Neutrino Mass and Mixing with Discrete Symmetry. *Rept. Prog. Phys.* **2013**, *76*, 056201. [CrossRef] [PubMed]
2. Pontecorvo, B. Neutrino Experiments and the Problem of Conservation of Leptonic Charge. *Sov. Phys. JETP* **1968**, *26*, 984. [Zh. Eksp. Teor. Fiz. 53 (1967) 1717].
3. Maki, Z.; Nakagawa, M.; Sakata, S. Remarks on the unified model of elementary particles. *Prog. Theor. Phys.* **1962**, *28*, 870. [CrossRef]
4. Lam, C.S. The Unique Horizontal Symmetry of Leptons. *Phys. Rev. D* **2008**, *78*, 073015. [CrossRef]
5. Lam, C.S. Determining Horizontal Symmetry from Neutrino Mixing. *Phys. Rev. Lett.* **2008**, *101*, 121602. [CrossRef] [PubMed]
6. Altarelli, G.; Feruglio, F. Discrete Flavor Symmetries and Models of Neutrino Mixing. *Rev. Mod. Phys.* **2010**, *82*, 2701. [CrossRef]
7. Kin, S.F.; Luhn, C. On the Origin of Neutrino Flavour Symmetry. *JHEP* **2009**, *0910*, 093.
8. King, S.F.; Merle, A.; Morisi, S.; Shimizu, Y.; Tanimoto, M. Neutrino Mass and Mixing: From Theory to Experiment. *New J. Phys.* **2014**, *16*, 045018. [CrossRef]
9. Machado, A.C.B.; Montero, J.C.; Pleitez, V. Three-Higgs-Doublet Model with A_4 Symmetry. *Phys. Lett. B* **2011**, *697*, 318. [CrossRef]
10. Branco, G.C.; Ferreira, P.M.; Lavoura, L.; Rebelo, M.N.; Sher, M.; Silva, J.P. Theory and Phenomenology of Two-Higgs-Doublet Models. *Phys. Rept.* **2012**, *516*, 1. [CrossRef]
11. Chaber, P.; Dziewit, B.; Holeczek, J.; Richter, M.; Zrałek, M.; Zajac, S. Lepton Masses and Mixing in a Two-Higgs-Doublet Model. *Phys. Rev. D* **2018**, *98*, 055007. [CrossRef]

12. Grimus, W.; Ludl, P.O. Finite Flavour Groups of Fermions. *J. Phys. A* **2012**, *45*, 233001. [CrossRef]
13. The GAP Group. GAP—Groups, Algorithms, and Programming; Version 4.7.6. 2014. Available online: http://www.gap-system.org/ (accessed on 1 January 2020).
14. Besche, H.U.; Eick, B.; O'Brien, E., The Small Groups Library; Version 2.1; (GAP 4.7.6 component). 2014. Available online: http://www.icm.tu-bs.de/ag_algebra/software/small/ (accessed on 1 January 2020).
15. Dabbaghian, V. Repsn, A GAP4 Package for Constructing Representations of Finite Groups; Version 3.0.2; (Refereed GAP package). 2011. Available online: http://www.sfu.ca/~vdabbagh/gap/repsn.html (accessed on 1 January 2020).

© 2020 by the authors. Licensee MDPI, Basel, Switzerland. This article is an open access article distributed under the terms and conditions of the Creative Commons Attribution (CC BY) license (http://creativecommons.org/licenses/by/4.0/).

Article

Doubly Charged Higgs Bosons and Spontaneous Symmetry Breaking at eV and TeV Scales

Janusz Gluza [1,2], Magdalena Kordiaczyńska [1,2,*] and Tripurari Srivastava [3,4]

1. Institute of Physics, University of Silesia, 40-007 Katowice, Poland; janusz.gluza@us.edu.pl
2. Faculty of Science, University of Hradec Králové, 500 03 Hradec Králové, Czech Republic
3. Department of Physics, Indian Institute of Technology, Kanpur 208016, India; tripurari@prl.res.in
4. Theoretical Physics Division, Physical Research Laboratory, Ahmedabad 380009, India
* Correspondence: mkordiaczynska@us.edu.pl

Received: 1 December 2019; Accepted: 7 January 2020; Published: 11 January 2020

Abstract: In this paper, beyond standard models are considered with additional scalar triplets without modification of the gauge group (Higgs Triplet Model—HTM) and with an extended gauge group $SU(2)_R \otimes SU(2)_L \otimes U(1)$ (Left–Right Symmetric Model—LRSM). These models differ drastically in possible triplet vacuum expectation values (VEV). Within the HTM, we needed to keep the triplet VEV at most within the range of GeV to keep the electroweak ρ parameter strictly close to 1, down to electronvolts due to the low energy constraints on lepton flavor-violating processes and neutrino oscillation parameters. For LRSM, the scale connected with the $SU(2)_R$ triplet is relevant, and to provide proper masses of non-standard gauge bosons, VEV should at least be at the TeV level. Both models predict the existence of doubly charged scalar particles. In this paper, their production in the e^+e^- collider is examined for making a distinction in the s- and t- channels between the two models in scenarios when masses of doubly charged scalars are the same.

Keywords: theoretical physics; particle physics; beyond Standard Model; scalar sector

1. Introduction

In 2012, the discovery of the neutral scalar particle, called the Higgs boson by the ATLAS [1] and CMS [2] collaborations, confirmed the mechanism of mass generation in the Standard Model (SM). However, SM can be an effective theory, similarly to as it was in the past with the Fermi four-interactions theory, and in particular, the scalar sector of the ultimate theory of elementary particles interactions may be more complex. One of the prime goals of Beyond Standard Model theories is a deeper understanding of neutrino tiny mass generation. An additional scalar triplet can explain the smallness of neutrino masses via the Type II seesaw mechanism. Also, there is a long-standing excess observed in anomalous magnetic moments of the muon which, if confirmed, cannot be explained within the Standard Model [3]. Furthermore, there are other phenomena in the universe which cannot be explained by the SM, that is, dark matter, baryon asymmetry, and dark energy, which call for the SM extensions and collider studies of possible exotic signals [4,5]. Extended scalar sectors include additional neutral and charged scalar particles. There are two ways to extend the scalar sector of the theory: directly, adding scalar fields, or indirectly, by extending the SM gauge group, which demands proper adjusting of the scalar sector. These additional particles can generate various lepton flavor and number violating processes, thus leaving signatures in the experiments. There are two facts which make them worth studying at colliders. Firstly, doubly charged scalars can produce the same sign dilepton signals at the colliders. Secondly, they are components of the triplet multiplets, an attractive scenario to explain neutrino masses. We focus on two popular models containing doubly charged scalars—SM with one extra triplet multiplet (HTM) and the left–right symmetric model (LRSM). These two models exhibit two very different scales of spontaneous symmetry breaking scales—eV

(HTM) and TeV (LRSM). Interestingly, though phenomenologically completely different, they can produce the same type of signatures at colliders when doubly charged scalars interact with leptons.

In this work, we focus on doubly charged scalar particle production in e^+e^- collisions within HTM and LRSM. Our main goal is to initialize the work in order to understand how both models can be differentiated when the doubly charged scalar $H^{\pm\pm}$ would be discovered. We discuss in detail relevant parameters of the model, carefully establishing benchmark points which give the same masses of $H^{\pm\pm}$ in both models, and analyze allowed scenarios for $H^{\pm\pm}$ decay branching ratios and possible $H^{\pm\pm}$ pair production in e^+e^- colliders.

1.1. Theoretical Introduction to HTM and LRSM

1.1.1. HTM

The HTM is one of the simplest extensions of the Standard Model. This model is based on the SM gauge group $SU(3)_C \times SU(2)_L \times U(1)_Y$. An additional $SU(2)_L$ scalar triplet is introduced in the particle content. We are following the convention $Y = 2Q - T_3$, where Q denotes the charge and T_3 is the third component of the isospin of the triplet. Depending on the hypercharge Y, the triplet contains neutral, singly- and doubly-charged scalars. In the studied case, the HTM's scalar sector is built of one scalar $SU(2)_L$ doublet Φ (in which the SM Higgs boson is situated) and the triplet Δ with $Y = 2$:

$$\Phi = \frac{1}{\sqrt{2}}\begin{pmatrix} \sqrt{2}w_\Phi^+ \\ v_\Phi + h_\Phi + iz_\Phi \end{pmatrix}, \quad \Delta = \frac{1}{\sqrt{2}}\begin{pmatrix} w_\Delta^+ & \sqrt{2}\delta^{++} \\ v_\Delta + h_\Delta + iz_\Delta & -w_\Delta^+ \end{pmatrix}. \tag{1}$$

where v_Φ and v_Δ denotes corresponding VEV, while w, h, z, δ are unphysical scalar fields. The physical singly charged state can be expressed by a combination of the doublet and triplet fields, where the doubly charged scalar field is already physical (see the Equations (A3a) and (A3c) in the Appendix A.1):

$$\begin{aligned} H^\pm &= -\sin\beta\, w_\Phi^\pm + \cos\beta\, w_\Delta^\pm, \quad \tan\beta = \frac{\sqrt{2}v_\Delta}{v_\Phi}, \\ H^{\pm\pm} &= \delta^{\pm\pm}. \end{aligned} \tag{2}$$

The doublet and triplet VEVs v_Φ and v_Δ are bounded by the condition:

$$v = \sqrt{v_\Phi^2 + 2v_\Delta^2} \simeq 246 \text{ GeV}, \tag{3}$$

where v is the SM electroweak symmetry breaking scale.

In Appendix A.1, the other physical fields and their masses are presented. From those considerations and from the decay $h \to \gamma\gamma$, it is known that the $|M_{H^{\pm\pm}} - M_{H^\pm}|$ mass gap does not overstep $\sim$50 GeV. That conclusion will be important for calculating the $H^{\pm\pm}$ decays and branching ratios.

In the HTM, we do not introduce the right-handed neutrino fields. Neutrinos get masses due to the Type II see-saw mechanism. As in the SM, left-handed leptons form doublets:

$$L_\ell = \begin{pmatrix} \nu_\ell \\ \ell \end{pmatrix}_L, \quad [\ell = e, \mu, \tau]. \tag{4}$$

Apart from the scalar potential presented in Appendix A.1, the Yukawa part of the Lagrangian should be added:

$$\mathcal{L}_Y^\Delta = \frac{1}{2}h_{\ell\ell'}L_\ell^T C^{-1} i\sigma_2 \Delta L_{\ell'} + \text{h.c.} \tag{5}$$

where C is the charge conjugation operator and $h_{\ell\ell'}$ is the symmetric Yukawa matrix. The Yukawa lagrangian L_Y^Δ provides massive neutrinos and interaction between triplet fields and leptons, particularly the $H^{\pm\pm} - l^{\mp} - l'^{\mp}$ vertex. In this case, the Yukawa coupling [6] reads:

$$Y_{\ell\ell'}^{HTM} = \frac{1}{\sqrt{2}v_\Delta} V_{PMNS}^* \, \mathrm{diag}\{m_1, m_2, m_3\} \, V_{PMNS}^\dagger, \qquad (6)$$

where m_i denotes neutrino masses, and the PMNS matrix V_{PMNS} is parametrized as follows:

$$V_{PMNS} = \begin{bmatrix} c_{12}c_{13}e^{i\alpha_1} & s_{12}c_{13}e^{i\alpha_2} & s_{13}e^{-i\delta_{CP}} \\ (-s_{12}c_{23} - c_{12}s_{23}s_{13}e^{i\delta_{CP}})e^{i\alpha_1} & (c_{12}c_{23} - s_{12}s_{23}s_{13}e^{i\delta_{CP}})e^{i\alpha_2} & s_{23}c_{13} \\ (s_{12}s_{23} - c_{12}c_{23}s_{13}e^{i\delta_{CP}})e^{i\alpha_1} & (-c_{12}s_{23} - s_{12}c_{23}s_{13}e^{i\delta_{CP}})e^{i\alpha_2} & c_{23}c_{13} \end{bmatrix}. \qquad (7)$$

This vertex depends on the neutrino parameters, where introducing neutrino masses directly breaks the lepton number. In Section 1.2, we will analyze the impact of this vertex and the doubly charged scalar's contribution to the lepton-number-violating (LNV) and lepton-flavor-violating (LFV) processes. Note that this coupling is also inversely proportional to the triplet VEV v_Δ, so the constraint coming from LFV and LNV processes will also bound v_Δ.

1.1.2. LRSM

In the case of LRSM, the gauge group is extended by the $\mathbf{SU(2)_R}$ right-handed group, where it is now $SU(3)_C \otimes \mathbf{SU(2)_R} \otimes SU(2)_L \otimes U(1)_Y$ [7,8]. There are a few ways to break this gauge symmetry down to the electroweak scale. For that, new scalar multiplets are introduced. We are following [7–9] with the scalar sector constructed of one bidoublet and two triplets with $Y = 2$, one under the $SU(2)_L$ and one under the $SU(2)_R$ group:

$$\Delta_{L,R} = \frac{1}{\sqrt{2}} \begin{pmatrix} w^+_{\Delta_{L,R}} & \sqrt{2}\delta^{++}_{L,R} \\ v_{\Delta_{L,R}} + h_{\Delta_{L,R}} + iz_{\Delta_{L,R}} & -w^+_{\Delta_{L,R}} \end{pmatrix}. \qquad (8)$$

Again, doubly charged scalar fields are obtained, which are already physical:

$$\begin{aligned} H_1^{++} &= \delta_L^{++}, \\ H_2^{++} &= \delta_R^{++}. \end{aligned} \qquad (9)$$

For the decomposition of singly-charged and neutral ones, see [9]. The whole scalar sector consists of two types of doubly charged scalar particles, two singly charged scalars, three neutral scalars (apart from the SM Higgs particle), and two pseudoscalars. Their masses are analyzed in the Appendix A.2. The LRSM realises the Type I See-Saw Mechanism. Additional right-handed neutrino fields are also present here, and form both left- and right-handed lepton doublet multiplets under the $SU(2)_L$ and $SU(2)_R$ group, respectively:

$$L_{iL} = \begin{pmatrix} \nu'_i \\ l'_i \end{pmatrix}_L, \quad L_{iR} = \begin{pmatrix} \nu'_i \\ l'_i \end{pmatrix}_R. \qquad (10)$$

This time, the Yukawa Lagrangian contains contributions from Δ_L and Δ_R [9]:

$$L_Y^l = -\bar{L}_R^c i\sigma_2 \Delta_L h_M L_L - \bar{L}_R^c i\sigma_2 \Delta_R h_M L_L + h.c., \qquad (11)$$

and again, the $H_{1,2}^{\pm\pm} - l^{\mp} - l^{\mp}$ vertex depends on heavy neutrino states masses and mixing. Since no explicit data for the mixing parameters of heavy neutrinos exists, we safely assumed that their couplings were diagonal (possible mixings are negligible for our scalar boson studies):

$$\mathcal{Y}_{\ell\ell'}^{\text{LRSM}} = \frac{1}{\sqrt{2}v_R} \text{diag}\{M_1, M_2, M_3\}. \qquad (12)$$

1.2. LFV Bounds on the Triplets VEV

The $H^{\pm\pm} - l^{\mp} - l'^{\mp}$ vertex contributes to many LFV and LNV processes. In the Table 1 we present the most relevant processes with corresponding experimental limits. The theoretical formulas for branching ratios for low-energy processes used in this publication are [10]:

Table 1. Low-energy LFV processes with $H^{\pm\pm}$ mediation and corresponding experimental limits.

Process:	Diagrams:	Limits:
Radiative decay: $l \to l'\gamma$		$\text{BR}(\mu \to e\gamma) \leq 4.2 \times 10^{-13}$ [11] $\text{BR}(\tau \to e\gamma) \leq 3.3 \times 10^{-8}$ [12] $\text{BR}(\tau \to \mu\gamma) \leq 4.4 \times 10^{-8}$ [12]
Three body decay: $l \to l_1 l_2 l_3$		$\text{BR}(\mu \to eee) \leq 1.0 \times 10^{-13}$ [13] $\text{BR}(\tau \to l_1 l_2 l_3) \leq \sim 10^{-8}$ [14]
$\mu \to e$ conversion: $\mu N \to eN^*$		$R(\mu^- \text{Au} \to e^- \text{Au}) \leq 7.0 \times 10^{-13}$ [15]

$$\begin{aligned}
\text{BR}(\mu \to e\gamma) &= \frac{\alpha_{em}}{192\pi} \frac{|(\mathcal{Y}^\dagger \mathcal{Y})_{e\mu}|^2}{G_F^2} \left(\frac{1}{M_{H^\pm}^2} + \frac{8}{M_{H^{\pm\pm}}^2}\right)^2 \text{BR}(\mu \to e\bar{\nu}_e \nu_\mu), \\
\text{BR}(\tau \to e\gamma) &= \frac{\alpha_{em}}{192\pi} \frac{|(\mathcal{Y}^\dagger \mathcal{Y})_{e\tau}|^2}{G_F^2} \left(\frac{1}{M_{H^\pm}^2} + \frac{8}{M_{H^{\pm\pm}}^2}\right)^2 \text{BR}(\tau \to e\bar{\nu}_e \nu_\tau), \\
\text{BR}(\tau \to \mu\gamma) &= \frac{\alpha_{em}}{192\pi} \frac{|(\mathcal{Y}^\dagger \mathcal{Y})_{\mu\tau}|^2}{G_F^2} \left(\frac{1}{M_{H^\pm}^2} + \frac{8}{M_{H^{\pm\pm}}^2}\right)^2 \text{BR}(\tau \to \mu\bar{\nu}_\mu \nu_\tau), \\
\text{BR}(\mu \to eee) &= \frac{1}{4G_F^2} \frac{|(\mathcal{Y}^\dagger)_{ee}(\mathcal{Y})_{\mu e}|^2}{M_{H^{\pm\pm}}^4} \text{BR}(\mu \to e\bar{\nu}\nu), \\
\text{BR}(\tau \to l_i l_j l_k) &= \frac{S}{4G_F^2} \frac{|(\mathcal{Y}^\dagger)_{\tau i}(\mathcal{Y})_{jk}|^2}{M_{H^{\pm\pm}}^4} \text{BR}(\tau \to \mu\bar{\nu}\nu), \quad S = \begin{cases} 1 & \text{if } j=k \\ 2 & \text{if } j \neq k \end{cases}.
\end{aligned} \qquad (13)$$

The rate of μ to e conversion in atomic nuclei [10,16,17] (Z_{eff}, Γ_{capt} and $F(q^2 \simeq -m_\mu^2)$ for the different atomic nuclei can be found in [18]):

$$R(\mu N \to eN^*) = \frac{(\alpha_{em} m_\mu)^5 Z_{\text{eff}}^4 Z |F(q)|^2}{4\pi^4 m_{\Delta^{\pm\pm}}^4 \Gamma_{\text{capt}}} \times \left| \frac{\mathcal{Y}_{e\mu}^\dagger \mathcal{Y}_{\mu\mu} F(r, s_\mu)}{3} - \frac{3(\mathcal{Y}^\dagger \mathcal{Y})_{e\mu}}{8} \right|^2$$

$$F(r, s_\mu) = \ln s_\mu + \frac{4s_\mu}{r} + \left(1 - \frac{2s_\mu}{r}\right) \times \sqrt{\left(1 + \frac{4s_\mu}{r}\right)} \ln \frac{\sqrt{(1+\frac{4s_\mu}{r})}+1}{\sqrt{(1+\frac{4s_\mu}{r})}-1}, \qquad (14)$$

$$r = -\frac{q^2}{m_{\Delta^{\pm\pm}}^2}, \quad s_\mu = \frac{m_\mu^2}{m_{\Delta^{\pm\pm}}^2}.$$

The $H^{\pm\pm}$ and $H^{\pm}$ contribution to the $(g-2)_\mu$ process are presented by diagrams in Figure 1. The analytical formulas can be found in Equations (15) (Diagram I) and (16) (Diagrams II and III) [19,20]. By $q_{l/H}$ we denote the lepton/scalar charge, where $m_{l/H}$ is the mass of the lepton/scalar particle.

$$[\Delta a_\mu]_I = -q_l \frac{m_\mu^2 |\mathcal{Y}_{\mu l}|^2}{8\pi^2} \int_0^1 dx \left[\frac{\left\{ x^2 - x^3 + \frac{m_l}{m_\mu} x^2 \right\}}{\left(m_\mu^2 x^2 + (m_l^2 - m_\mu^2)x + M_H^2(1-x) \right)} \right], \tag{15}$$

$$[\Delta a_\mu]_{II,\,III} = -q_H \frac{m_\mu^2 |\mathcal{Y}_{\mu l}|^2}{8\pi^2} \int_0^1 dx \left[\frac{\left\{ (x^3 - x^2) + \frac{m_l}{m_\mu}(x^2 - x) \right\}}{\left(m_\mu^2 x^2 + (M_H^2 - m_\mu^2)x + (1-x)m_l^2 \right)} \right]. \tag{16}$$

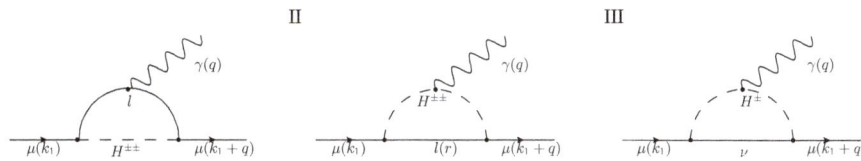

Figure 1. Singly- and doubly-charged scalars' contributions to $(g-2)_\mu$.

2. A Case Study: Benchmark Points for the Scenario With $m_{H^{\pm\pm}} = 700$ GeV in HTM and LRSM

In this work, we examine the case when a mass of $H^{\pm\pm}$ is 700 GeV.

From the LHC data (see Figure 13 in [21]), to keep $M_{H^{\pm\pm}} \sim 700$ GeV, a doubly charged scalar's decays must satisfy the following conditions:

$$\begin{array}{ll} \text{BR}(H^{\pm\pm} \to ee) < 0.5 & \text{BR}(H^{\pm\pm} \to \mu\mu) < 0.3, \\ \text{BR}(H^{\pm\pm} \to e\mu) < 0.5 & \sum_{e,\mu} \text{BR}(H^{\pm\pm} \to ll') < 0.7. \end{array} \tag{17}$$

The above limits apply to the doubly charged scalar particles coupling to the left-handed leptons, but since in the LRSM we assume the degenerated case $M_{H_1^{\pm\pm}} = M_{H_2^{\pm\pm}}$, and the couplings $H_{1,2}^{\pm\pm} - l - l$ are equal, $H_2^{\pm\pm}$ has to fulfil the same condition. Those limits were calculated assuming that the doubly charged scalar particles decay 100% to leptons. In Section 3.2, we show that this assumption is justified both for HTM and LRSM. Relative leptonic branching ratios do not depend on the triplet VEV, but vary depending on neutrino parameters, as well as Majorana phases (see Figure 1 in [22]). For further calculation, we used the neutrino oscillation data from [23,24] within the 2σ confidence level range and Majorana phases $\alpha_{1,2} \in (0, 2\pi)$ (see the PMSN matrix parametrisation in Equation (7)).

In the case of LRSM, we covered the v_R region examined by the LHC with corresponding heavy neutrinos' masses [25,26]. Details are discussed in Section 3.1. Regarding the scalar particles' masses, we constructed a scalar mass spectrum in which $M_{H^{\pm\pm}} = 700$ GeV. Corresponding parameters of scalar potentials in both models are given in Table 2.

The mass benchmark points were constructed in order to satisfy several conditions.

For HTM, the potential stability imposed the following relation between the model parameters [27,28]:

$$\begin{array}{l} \lambda \geq 0, \; \lambda_2 + \frac{\lambda_3}{2} \geq 0, \; \lambda_1 + \sqrt{\lambda(\lambda_2 + \lambda_3)} \geq 0, \; \lambda_1 + \lambda_4 + \sqrt{\lambda(\lambda_2 + \lambda_3)} \geq 0, \\ |\lambda_4|\sqrt{\lambda_2 + \lambda_3} - \lambda_3 \sqrt{\lambda} \geq 0 \quad \text{or} \quad 2\lambda_1 + \lambda_4 + \sqrt{(2\lambda\lambda_3 - \lambda_4^2)(\frac{2\lambda_2}{\lambda_3} + 1)} \geq 0. \end{array} \tag{18}$$

On the other hand, from unitarity constraints, we got [29,30]:

$$\begin{array}{l} \text{Max}\Big\{ \left|\frac{\lambda}{2}\right|, \; |\lambda_1|, \; \frac{1}{2}|2\lambda_1 + 3\lambda_4|, \; |\lambda_1 + \lambda_4|, \; \frac{1}{2}|2\lambda_1 - \lambda_4|, \; |\lambda_3 - 2\lambda_2|, \; |2\lambda_2|, \\ |2(\lambda_3 + \lambda_2)|, \; \frac{1}{4}\left|3\lambda + 16\lambda_2 + 12\lambda_3 \pm \sqrt{(3\lambda - 16\lambda_2 - 12\lambda_3)^2 + 24(\lambda_4 + 2\lambda_1)^2}\right|, \\ \frac{1}{4}\left|\lambda + 4\lambda_2 + 8\lambda_3 \pm \sqrt{(\lambda - 4\lambda_2 - 8\lambda_3)^2 + 16\lambda_4^2}\right| \Big\} \geq 16\pi \; (8\pi). \end{array} \tag{19}$$

Table 2. Exemplary benchmark points and corresponding potential parameters for HTM ($v_\Delta = 15$ eV) and LRSM ($v_R = 7000$ GeV) with $M_{H^{\pm\pm}_{1,2}} = 700$ GeV. The scalar potential parameters are defined in the Appendix A.1, Equations (A1) and (A6).

HTM	$\mu = 1.72 \times 10^{-7}$ $\lambda = 0.519$ $\lambda_1 = 0.519$ $\lambda_2 = 16.7$ $\lambda_3 = 0.$ $\lambda_4 = 0.$ $M_h = 125$ $M_H = 700$ $M_{H^\pm} = 700$ $M_{H^{\pm\pm}} = 700$
LRSM	$\lambda_1 = 0.129$ $\rho_1 = 0.00375$ $\rho_2 = 0.00375$ $\rho_3 - 2\rho_1 = 0.015$ $\alpha_3 = 4.0816$ $2\lambda_2 - \lambda_3 = 0$ $M_{H^0_0} = 125$ $M_{H^0_1} = 10\,000$ $M_{H^0_2} = 606$ $M_{H^0_3} = 606$ $M_{H^{\pm\pm}_1} = 700$ $M_{H^{\pm\pm}_2} = 700$ $M_{H^\pm_1} = 655$ $M_{H^\pm_2} = 10\,003$

The right side of the above inequality depends on the convention, whether we chose the scattering matrix element $\mathcal{M}$ less than 16π (what corresponds with the 0th partial wave amplitude $|a_0| \leq 1$, see Equation (1) in [30]) or 8π ($|\text{Re}(a_0)| \leq \frac{1}{2}$). Figure 2 in Reference [27] presents the mass region plot for heavy neutral scalar M_H and singly charged scalar $M_{H^\pm}$ allowed from potential stability, unitarity, and the T parameter [31,32] for the triplet VEV 1 eV. The calculations for $M_{H^\pm}$ and $M_{H^{\pm\pm}}$ in Figure 2 using both $\mathcal{M} < 16\pi$ and $\mathcal{M} < 8\pi$ constrained the maximum $|M_{H^{\pm\pm}} - M_{H^\pm}|$ gap, and thanks to that, we could determine possible $H^{\pm\pm}$ decay channels. The mass gap should be less than M_W. Taking those results into account, we found our choice of a degenerate mass case $M_{H^{\pm\pm}} = M_{H^\pm} = M_H$ fulfilled the potential stability, unitarity, and the T parameter restriction and bounds from $h \to \gamma\gamma$ [27,32–34].

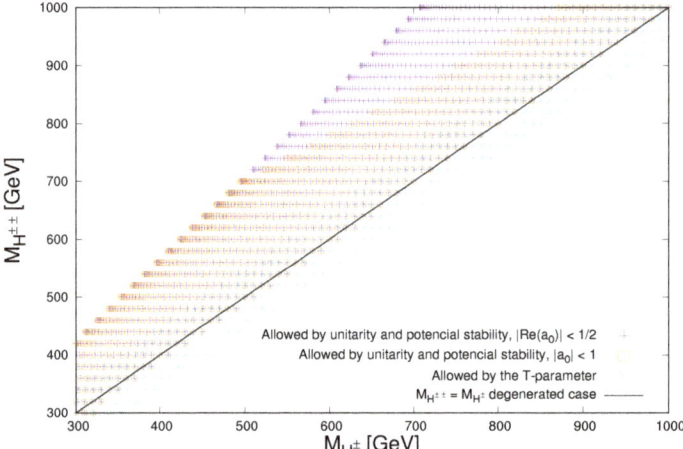

Figure 2. Singly and doubly charged scalars' mass dependence with limits coming from unitarity, potential stability, and the T parameter for $v_\Delta = 1$ eV.

3. Results and Discussion

In this section, we will apply determined parameters of the HTM and LRSM models and calculate possible VEV scales, and in order to compare the two models, the $H^{\pm\pm}_{(1,2)}$ branching ratios. In numerical calculations, fortran and Mathematica scripts were used, and the cross-section $e^+e^- \to H^{\pm\pm}_{(1,2)} H^{\mp\mp}_{(1,2)}$ process was calculated using MadGraph [35].

3.1. Limits on the Triplet VEV

The limits on the triplet VEVs come from several sources. Firstly, the additional scalar triplets under the $SU(2)_L$ group impact the value of the ρ parameter which relates masses of W and Z bosons with gauge SM gauge couplings [36] or gives a ratio of charged and neutral currents (see the Appendix in [37]). Taking experimental data into account, $\rho^{exp} = 1.00037 \pm 0.00023$, scalar triplet v_Δ is restricted from above, and the maximum VEV is of the order of 1.7 GeV. We assumed that for LRSM, the VEV of the $SU(2)_L$ scalar triplet is equal to zero. That choice allows for avoidance of the fine-tuning problem discussed in [8,38]. In HTM, the $SU(2)_L$ scalar triplet is restricted from the bottom by the low energy constraints, where it is not possible to set it to zero. In Table 3, we present the lowest limit on the HTM triplet VEV for $M_{H^{\pm\pm}} = 700$ GeV. The limits come from solutions to the relations in Equations (6) and (7), as well as experimental data on masses and mixing of neutrinos.

Table 3. Lower limits on the triplet vacuum expectation value v_Δ (in eV) for doubly charged scalar's mass $M_{H^{\pm\pm}} = 700$ GeV. We calculated the above results by scanning through the entire space of neutrino oscillation parameters, that is, within a $\pm 2\sigma$ confidence level range of mixing parameters [23,24], as well as the whole range of Majorana phases $\alpha_{1,2}$, taking into account the cosmological neutrino mass limit $\sum_{i=1}^{3} m_{\nu_i} < 0.23$ eV [39].

	NH	IH
min v_Δ [eV]	0.93	1.07

In the case of the $SU(2)_R$ scalar triplet VEV in LRSM, the ρ parameter was preserved if $v_R \gg \kappa_+$, where κ_+ is the SM electroweak symmetry breaking scale (see Equation (42) in [9]). On the other hand, to provide correct masses of heavy scalars and right-handed gauge bosons, v_R had to be at least at the TeV range. As we were interested in the region potentially examined by the LHC, we needed to restrict its value to the range $v_R \in 10^3 \div 10^4$ GeV. As the LNV and LFV bounds discussed in Section 1.2 depend on heavy neutrino masses, using the relation between the heavily charged gauge boson's mass and $SU(2)_R$ triplet VEV:

$$M_{W_2}^2 \simeq \frac{g^2 v_R^2}{2} \quad \Rightarrow \quad M_{W_2} \simeq 0.47\, v_R. \qquad (20)$$

We were able to find the parameter space for the triplet VEV v_R and heavy neutrino masses. For that, we used the CMS experimental data from the $pp \to lljj$ process. Figure 6 in [25] and Figure 7 in [26] present $M_{W_2} - M_N$ exclusion plots, assuming $M_{W_2} > M_N$. For convenience, we repeat them here (see Figure 3). We used these data and exclusion plots for further analysis.

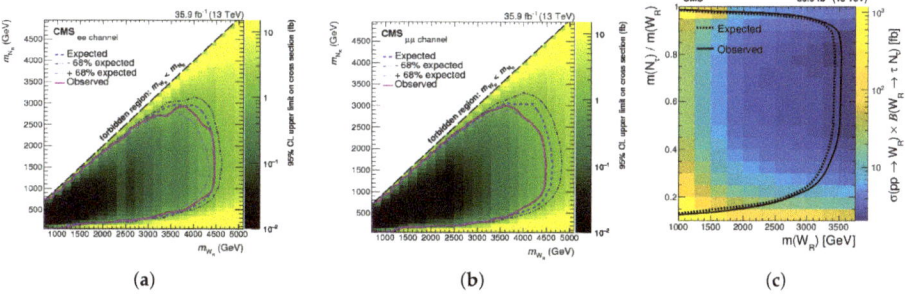

Figure 3. Upper limit on the $pp \to lljj$ cross-section for different and mass hypotheses, for the electron (**a**), muon (**b**), and taon (**c**) channels. The thin-dotted (blue) curves in the Figures (**a**) and (**b**) indicate the region in (M_{W_2}, M_{N_i}) parameter space that is expected to be excluded at 68% CL [25,26].

3.2. Doubly Charged Scalar Particles Decays within the LRSM and HTM

3.2.1. HTM

In the HTM model, the doubly charged scalar can decay to leptons, a $W^{\pm}$ boson, and singly charged scalar particles. Because the singly and doubly charged scalars' mass gap is less than M_W, the $H^{\pm\pm} \to H^{\pm}H^{\pm}$ and $H^{\pm\pm} \to H^{\pm}W^{\pm}$ processes are suppressed. There are two possibilities left: $H^{\pm\pm} \to l_i^{\pm}l_j^{\pm}$ and $H^{\pm\pm} \to W^{\pm}W^{\pm}$. Since $H^{\pm\pm} - W^{\pm} - W^{\pm}$ is proportional to the triplet VEV v_Δ, and $H^{\pm\pm} - l_i^{\pm} - l_j^{\pm} \sim v_\Delta^{-1}$, leptonic channels dominate, from 0 up to $v_\Delta \sim 10^5$ eV (see Figure 4 in [40]), so our assumption that doubly charged particles decay purely to leptons is valid for our benchmark points, especially when we are interested in the low v_Δ values at the range of $1 \div 10$ eV.

3.2.2. LRSM

In the LRSM, the doubly charged scalar particles can decay to gauge bosons W_1, W_2, and singly charged scalars $H_1^{\pm}$ and $H_2^{\pm}$. Some of those decays are connected with vertices which are proportional to the $SU(2)_L$ triplet VEV $v_L = 0$ or to the $W_1 - W_2$ mixing angle $\zeta \lesssim 10^{-2}$ [41,42]. Also, as $M_{H_{1,2}^{\pm\pm}} \ll M_{H_1^{\pm}}$ (see Equation (A8h)) and $M_{H_{1,2}^{\pm\pm}} \ll M_{W_2}$, diagrams involving heavy gauge bosons are suppressed for the 1.5 TeV e^+e^- collision energy, again, leptonic decays of doubly charged scalar particles dominate.

3.3. Doubly Charged Scalar Particles' Pair Production at Future High Energies

Let us consider the potential production of the doubly charged scalar particle pair at a e^+e^- collision energy of 1.5 TeV. A TeV energy range of e^+e^- coliders has been studied intensively in the past. Presently, the only considered scenario with such extreme collision energies of leptons is the CLIC project [43] (in future, extreme energies may become possible in Plasma Wakefield Linear Colliders [44]). The list of Feynman diagrams is given in Figure 4. In the s-channel, a $H^{\pm\pm}$ pair production goes by scalar and gauge bosons (Figure 4a), and in the t-channel, production is connected with an exchange of the charged lepton (Figure 4b).

For the s-channel, in two considered models, the photon, Z boson, and SM-like Higgs particles contribute. In HTM, additional scalar H can also couple to leptons and doubly charged scalar particles, where in LRSM, H_1^0 and H_2^0 also contribute. Examining couplings carefully, we can find that the scalars' contribution to this process is negligible. In LRSM the heavy gauge Z_2 boson is also present in the s-channel.

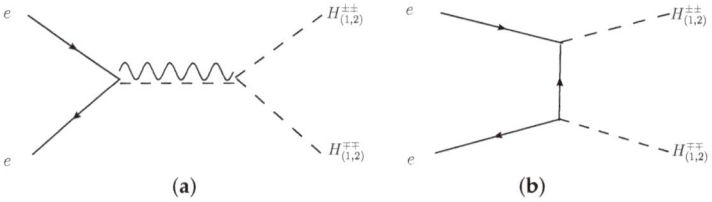

Figure 4. Feynman diagrams for the (**a**) s-channel and (**b**) t-channel pair production of doubly charged scalar particles in the e^+e^- collision. The following particles contribute to the diagrams: HTM: γ, Z, h and H (s-channel); e, μ, τ (t-channel). LRSM: γ, Z_1, Z_2, H_0^0, H_1^0, H_2^0 (s-channel); e (t-channel). The LFV breaking vertex $H_{1,2}^{\pm\pm} - l_i - l_j$ is not present since Yukawa couplings are assumed to be diagonal (see Equation (12)).

The t-channel diagram contains the $H^{\pm\pm}_{(1,2)} - l - l'$ vertex, which depends on the triplet VEV (see Equations (6) and (12)). Figure 5 presents the cross-section for the $e^+e^- \to H^{\pm\pm}_{(1,2)} H^{\mp\mp}_{(1,2)}$ process and its dependence on the triplet VEVs within the HTM (left) and LRSM (right) models.

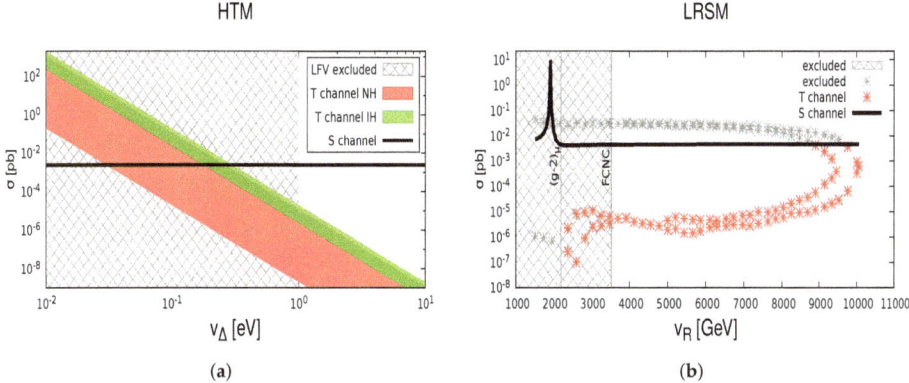

Figure 5. Doubly charged particles' pair production $e^+e^- \to H^{++}H^{--}$ for $M_{H^{\pm\pm}_{(1,2)}} = 700$ GeV and CM energy 1.5 TeV. (**a**) Left figure is for the HTM and (**b**) LRSM. The choice of the parameter space is discussed in Sections 2 and 3.1. The crossed area is excluded by (**a**) the LFV processes (**b**) and $(g-2)_\mu$ and FCNC. The maximum for $v_R = 1900$ GeV comes from the Z_2 resonance.

As expected, the t-channel dominates for the lower triplet VEV v_Δ. The shaded regions correspond to normal (red area) and inverted (green area) neutrino mass hierarchies, with neutrino parameters smeared within $\pm 2\sigma$ for Majorana phases $\alpha_{1,2} \in (0, 2\pi)$, taking into account the cosmological neutrino mass limit $\sum m_{\nu_i} < 0.23$ eV [39]. On the other hand, taking into account LFV processes and limits given in Table 3, we can see that triplet VEV $v_\Delta < 1$ eV is forbidden by the low energy experiments, so the whole region where the t-channel could bring a significant contribution is excluded.

In LRSM, we studied possible triplet VEV values using LHC CMS data for possible heavy neutrino masses N_e, N_μ and N_τ within the 68% confidence level, assuming $M_{W_2} > M_{N_i}$ (see Figure 3). The result is given in Figure 5b). We assumed diagonal Yukawa couplings and no LFV in the $H^{\pm\pm}_{1,2} - l - l$ vertex. Still, additional constraints came from experimental data which was discussed in Section 1.2, such as couplings $\mathcal{Y}_{ee}$ and $\mathcal{Y}_{\mu\mu}$ couplings which was a constraint by the $(g-2)_\mu$ experiment (see Equations (15) and (16)), $e^+e^- \to l^+l^-$ [45], and the Møller scattering [46]. Both branching ratios Equation (17) and FCNC limits on the scalar particles masses were also crucial in our analysis (see Appendix A.1).

Taking the above bounds into account, we excluded the triplet VEV below $\sim$3600 GeV and the parameter region where the t-channel could dominate over the s-channel. However, in contrast to the HTM model, the t-channel's contribution is still comparable with the s-channel, and cannot be neglected.

Regarding the s-channels, both in HTM and LRSM, this channel does not depend on the triplet VEV, and a significant Z_2 contribution is excluded by the $(g-2)_\mu$ and FCNC conditions, so a heavy gauge boson contribution does not affect the final results for the s-channel contributions and $H^{++}_{(1,2)} H^{--}_{(1,2)}$ pair production.

Here, we have focused on the $H^{\pm\pm}$ pair production process in e^+e^- collisions. As signals might be observable at some regions of model parameters, in future we plan to carefully study subsequent decays of $H^{\pm\pm}$, as well as background processes. It is worth noticing that doubly charged scalars can exhibit small decay widths, thus having long life-spans [47,48], and they do not leave any signatures in the detector-charged tracker, or can even escape the detector. However, they can deposit energy on different sub-detectors. Thus, they can be searched for through displaced secondary vertices analysis.

This kind of search can be important for feeble interactions of doubly charged scalars, and can be complementary to same-sign dilepton decay studies of doubly charged scalars with prompt decays. Studying such scenarios is also on our agenda. Searches for heavy long-lived multi-charged particles have already been initiated by many collider experiments, such as ATLAS, CMS, and Tevatron [49–52].

4. Summary and Outlook

In this article, we have discussed the present status of the simplest beyond-SM models which include doubly charged scalar particles. We presented the status of experimental data relevant for the determination of non-standard spontaneous symmetry breaking and VEVs in both models. We made a case study for a realistic scenario when a mass of this particle is 700 GeV, taking into account all relevant experimental constraints, and discussed its decay channels and a possibility of $H^{\pm\pm}$ pair production in future e^+e^- colliders, as a function of allowed VEV. In HTM, the low-energy experiments, such as $(g-2)_\mu$, ρ-parameter, and LFV and LNV processes put stringent constraints over triplet VEV such that $v_\Delta \lesssim \mathcal{O}$ (eV), whereas right-handed triplet VEV in LRSM was constrained by the search for the new charged gauge boson which required $v_R \gtrsim \mathcal{O}$ (TeV), depending on the mass of right-handed heavy neutrinos. We provided the benchmark for both the models considering the mass of a doubly charged scalar to be 700 GeV. In the proposed e^+e^- colliders, the doubly charged scalar can be produced in pairs via the Drell-Yan process, heavy neutral bosons, and BSM neutral scalars. In HTM, pair production is dominated by the t-channel in region $v_\Delta \gtrsim \mathcal{O}$ (eV), which is excluded by the low energy experiments. In LRSM, the right-handed breaking scale $v_R < 3600$ GeV is excluded by low energy constraints, including FCNC, but the t-channel can still be comparable with the s-channel, and provides appreciable contribution to the doubly charged scalar pair production, in contrast to HTM. The s-channel contribution to the pair production process is, however, practically the same in both models for the considered realistic benchmarks. Therefore, the conclusion is that the signals are more promising in the case of LRSM, where the t-channel can be comparable with s-channel contributions for v_R values up to 10 TeV. We plan to study such cases more carefully in the future, paying more attention to the decay modes, background processes, and scenarios with displaced vertices. However, a more relevant option in the context of our studies seems to be pp colliders, HL-LHC, and particularly, the FCC-hh collider, which opens up the window for huge energies of proton–proton collisions up to 100 TeV [4,5,53].

Author Contributions: Conceptualization, investigation, methodology, resources: J.G., M.K. and T.S.; software, validation: M.K. and T.S.; writing, editing: J.G. and M.K.; visualization: M.K. and T.S.; supervision, project administration: J.G.; funding acquisition: J.G. and M.K. All authors have read and agreed to the published version of the manuscript.

Funding: This research was funded by the Polish National Science Center (NCN) under grant 2015/17/N/ST2/04067, the international mobilities for research activities of the University of Hradec Králové, CZ.02.2.69/0.0/0.0/16 027/0008487 and by COST (European Cooperation in Science and Technology) Action CA16201 PARTICLEFACE.

Acknowledgments: We thank Joydeep Chakrabortty and Dipankar Das for useful discussions.

Conflicts of Interest: The authors declare no conflict of interest.

Abbreviations

The following abbreviations are used in this manuscript:

SM	Standard Model
HTM	Higgs Triplet Model
LRSM	left–right Symmetric Model
VEV	Vacuum Expectation Value
NH	Normal Hierarchy of neutrino masses
IH	Inverted Hierarchy of neutrino masses
LNV	Lepton Number Violation
LFV	Lepton flavor Violation

Appendix A

Appendix A.1. Scalar Particles within the HTM

The most general scalar potential with an additional triplet has the following form [29]:

$$V = -m_\Phi^2 (\Phi^\dagger \Phi) + \tfrac{\lambda}{4} (\Phi^\dagger \Phi)^2 + M_\Delta^2 \operatorname{Tr}(\Delta^\dagger \Delta) + \left[\mu \left(\Phi^T i\sigma_2 \Delta^\dagger \Phi\right) + \text{h.c.}\right] + \\ + \lambda_1 (\Phi^\dagger \Phi) \operatorname{Tr}(\Delta^\dagger \Delta) + \lambda_2 \left[\operatorname{Tr}(\Delta^\dagger \Delta)\right]^2 + \lambda_3 \operatorname{Tr}\left[(\Delta^\dagger \Delta)^2\right] + \lambda_4 \Phi^\dagger \Delta \Delta^\dagger \Phi, \quad \text{(A1)}$$

where:

$$m_\Phi^2 = \frac{\lambda}{4} v_\Phi^2 + \frac{(\lambda_1 + \lambda_4)}{2} v_\Delta^2 - \sqrt{2}\mu \, v_\Delta, \quad \text{(A2a)}$$

$$M_\Delta^2 = -(\lambda_2 + \lambda_3) v_\Delta^2 - \frac{(\lambda_1 + \lambda_4)}{2} v_\Phi^2 + \frac{\mu}{\sqrt{2}} \frac{v_\Phi^2}{v_\Delta}. \quad \text{(A2b)}$$

The physical fields and their masses are:

$$H^{\pm\pm} = \delta^{\pm\pm} \quad \text{(A3a)}$$

$$\begin{pmatrix} G_0 \\ A \end{pmatrix} = \begin{pmatrix} \cos\beta' & \sin\beta' \\ -\sin\beta' & \cos\beta' \end{pmatrix} \begin{pmatrix} z_\Phi \\ z_\Delta \end{pmatrix}, \quad \tan\beta' = \frac{2v_\Delta}{v_\Phi}, \quad \text{(A3b)}$$

$$\begin{pmatrix} G^\pm \\ H^\pm \end{pmatrix} = \begin{pmatrix} \cos\beta & \sin\beta \\ -\sin\beta & \cos\beta \end{pmatrix} \begin{pmatrix} w_\Phi^\pm \\ w_\Delta^\pm \end{pmatrix}, \quad \tan\beta = \frac{\sqrt{2}v_\Delta}{v_\Phi}, \quad \text{(A3c)}$$

$$\begin{pmatrix} h \\ H^0 \end{pmatrix} = \begin{pmatrix} \cos\alpha & \sin\alpha \\ -\sin\alpha & \cos\alpha \end{pmatrix} \begin{pmatrix} h_\Phi \\ h_\Delta \end{pmatrix}, \quad \tan 2\alpha = \frac{2B_S}{C_S - A_S}. \quad \text{(A3d)}$$

$$M_{H^\pm}^2 = \frac{(2\sqrt{2}\mu - \lambda_4 v_\Delta)}{4v_\Delta}(v_\Phi^2 + 2v_\Delta^2), \quad \text{(A4a)}$$

$$M_{H^{\pm\pm}}^2 = \frac{\mu v_\Phi^2}{\sqrt{2}v_\Delta} - \frac{\lambda_4}{2} v_\Phi^2 - \lambda_3 v_\Delta^2, \quad \text{(A4b)}$$

$$M_A^2 = \frac{\mu}{\sqrt{2}v_\Delta}(v_\Phi^2 + 4v_\Delta^2), \quad \text{(A4c)}$$

$$M_h^2 = \frac{1}{2}\left((A_S + C_S) - \sqrt{(A_S - C_S)^2 + 4B_S^2}\right), \quad \text{(A4d)}$$

$$M_{H^0}^2 = \frac{1}{2}\left((A_S + C_S) + \sqrt{(A_S - C_S)^2 + 4B_S^2}\right). \quad \text{(A4e)}$$

where:

$$A_S = \frac{\lambda v_\Phi^2}{2}, \quad \text{(A5a)}$$

$$B_S = \sqrt{2}\mu v_\Phi - (\lambda_1 + \lambda_4) v_\Delta v_\Phi, \quad \text{(A5b)}$$

$$C_S = \frac{\mu v_\Phi^2}{\sqrt{2} v_\Delta} + 2(\lambda_2 + \lambda_3) v_\Delta^2. \quad \text{(A5c)}$$

We are following the notation from [29]. We express the lagrangian coefficients λ, λ_i, μ (see Equation (A1)) as the functions of scalar particles' masses:

$$\lambda_1 = -\frac{2}{v_\Phi^2 + 4v_\Delta^2} M_A^2 + \frac{4}{v_\Phi^2 + 2v_\Delta^2} M_{H^\pm}^2 + \frac{\sin 2\alpha}{2v_\Phi v_\Delta}(M_h^2 - M_H^2), \qquad (A6a)$$

$$\lambda_2 = \frac{1}{v_\Delta^2}\left\{ \frac{s_\alpha^2 M_h^2 + c_\alpha^2 M_H^2}{2} + \frac{1}{2}\frac{v_\Phi^2}{v_\Phi^2 + 4v_\Delta^2} M_A^2 - \frac{2v_\Phi^2}{v_\Phi^2 + 2v_\Delta^2} M_{H^\pm}^2 + M_{H^{\pm\pm}}^2 \right\} \qquad (A6b)$$

$$\lambda_3 = \frac{1}{v_\Delta^2}\left\{ \frac{-v_\Phi^2}{v_\Phi^2 + 4v_\Delta^2} M_A^2 + \frac{2v_\Phi^2}{v_\Phi^2 + 2v_\Delta^2} M_{H^\pm}^2 - M_{H^{\pm\pm}}^2 \right\} \qquad (A6c)$$

$$\lambda_4 = \frac{4}{v_\Phi^2 + 4v_\Delta^2} M_A^2 - \frac{4}{v_\Phi^2 + 2v_\Delta^2} M_{H^\pm}^2 \qquad (A6d)$$

$$\lambda = \frac{2}{v_\Phi^2}\left\{ c_\alpha^2 M_h^2 + s_\alpha^2 M_H^2 \right\}, \qquad (A6e)$$

$$\mu = \frac{\sqrt{2} v_\Delta}{v_\Phi^2 + 4v_\Delta^2} M_A^2, \qquad (A6f)$$

where s_α and c_α denote $\sin\alpha$ and $\cos\alpha$ (Equations (A3d) and (A5)). We used an approximation $s_\alpha = \sin\alpha \sim 2\frac{v_\Delta}{v} \to 0$ [27]. Substituting the potential parameters in Equation (A4) by Equation (A6), relations between masses of physical states are obtained. They are not independent, and we chose $M_{H^{\pm\pm}}$, M_H and $M_h = 125$ GeV as external parameters for $M_{H^\pm}$ and M_A:

$$M_A = \sqrt{\frac{(v_\Phi^2 + 4v_\Delta^2)(M_H^2 c_\alpha^2 + M_h^2 s_\alpha^2 - v_\Delta^2)}{v_\Phi^2}}, \qquad (A7a)$$

$$M_{H^\pm} = \sqrt{\frac{\left(M_{H^{\pm\pm}}^2 + \frac{v_\Phi^2 M_A^2}{v_\Phi^2 + 4v_\Delta^2} - v_\Delta^2\right)(v_\Phi^2 + 2v_\Delta^2)}{2v_\Phi^2}} \qquad (A7b)$$

Appendix A.2. The Mass Spectrum in LRSM

The scalar potential for LRSM with one bidoublet and two triplets is given in Equation (25) in [9]. Scalar particles masses as functions of potential parameters are presented in Equation (A8).

$$M_{H_0^0}^2 \simeq 2\kappa_+^2 \lambda_1 \qquad = 125 \text{ GeV} \qquad (A8a)$$

$$M_{H_1^0}^2 \simeq \frac{1}{2}\alpha_3 v_R^2 \qquad \geq 10 \text{ TeV} \qquad (A8b)$$

$$M_{H_2^0}^2 \simeq 2\rho_1 v_R^2 \qquad (A8c)$$

$$M_{H_3^0}^2 \simeq \frac{1}{2} v_R^2 (\rho_3 - 2\rho_1) \qquad \geq 55.4 \text{ GeV} \qquad (A8d)$$

$$M_{A_1^0}^2 \simeq \frac{1}{2}\alpha_3^2 v_R^2 - 2\kappa_+^2(2\lambda_2 - \lambda_3) \qquad \geq 10 \text{ TeV} \qquad (A8e)$$

$$M_{A_2^0}^2 \simeq \frac{1}{2} v_R^2 (\rho_3 - 2\rho_1) \qquad (A8f)$$

$$M_{H_1^\pm}^2 \simeq \frac{1}{2} v_R^2 (\rho_3 - 2\rho_1) + \frac{1}{4}\alpha_3 \kappa_+^2 \qquad \geq 10 \text{ TeV} \qquad (A8g)$$

$$M_{H_2^\pm}^2 \simeq \frac{1}{2}\alpha_3 \left[v_R^2 + \frac{1}{2}\kappa_+^2\right] \qquad (A8h)$$

$$M_{H_1^{\pm\pm}}^2 \simeq \frac{1}{2}\left[v_R^2(\rho_3 - 2\rho_1) + \alpha_3 \kappa_+^2\right] \qquad (A8i)$$

$$M_{H_2^{\pm\pm}}^2 \simeq 2\rho_2 v_R^2 + \frac{1}{2}\alpha_3 \kappa_+^2 \qquad (A8j)$$

where $k_+ = 246$ GeV is a combination of bidoublet VEVs. H_0^0 corresponds with the SM Higgs boson, so it has a mass of 125 GeV. Neutral H_1^0 and A_1^0 particles intermediate the Flavor-Changing Neutral Current (FCNC) processes [54], so their masses (Equations (A8b) and (A8e)) must be higher than 10 TeV to suppress this effect. Some limits on the other neutral scalar particles comes from the LEP-II experiment $M_{H_3^0} \leq 55.4$ GeV (Equation (A8d)). This time, we cannot assume the mass degeneration, and from Equations (A8g)–(A8j) it is obvious that $M_{H_1^{\pm\pm}} \neq M_{H_1^\pm}$ and $M_{H_2^{\pm\pm}} \neq M_{H_2^\pm}$. For $M_{H_1^{\pm\pm}} = 700$ GeV maximum value of $M_{H_1^\pm}$ is equal to 655 GeV, so the minimum mass gap is less than the W_1 mass: $M_{H_1^{\pm\pm}} - M_{H_1^\pm} > 45$ GeV. The $M_{H_2^\pm}$ is greater than 10 TeV (to be compared with Equation (A8b)), so we will ignore the $H_2^{\pm\pm} \to H_2^\pm + X$ decay. The whole mass spectrum and corresponding potential parameters we used in this paper are shown in the Table 2. Those values also fulfill the potential stability and unitarity bounds (see Equations (5)–(10) and (21) in [55]).

References

1. Aad, G.; Abajyan, T.; Abbott, B.; Abdallah, J.; Khalek, S.A.; Abdelalim, A.A.; Abdinov, O.; Aben, R.; Abi, B.; Abolins, M.; et al. Observation of a new particle in the search for the Standard Model Higgs boson with the ATLAS detector at the LHC. *Phys. Lett.* **2012**, *B716*, 1–29. [CrossRef]
2. Chatrchyan, S.; Khachatryan, V.; Sirunyan, A.M.; Tumasyan, A.; Adam, W.; Aguilo, E.; Bergauer, T.; Dragicevic, M.; Erö, J.; Fabjan, C.; et al. Observation of a New Boson at a Mass of 125 GeV with the CMS Experiment at the LHC. *Phys. Lett.* **2012**, *B716*, 30–61. [CrossRef]
3. Campanario, F.; Czyż, H.; Gluza, J.; Jeliński, T.; Rodrigo, G.; Tracz, S.; Zhuridov, D. Standard model radiative corrections in the pion form factor measurements do not explain the a_μ anomaly. *Phys. Rev.* **2019**, *D100*, 076004. [CrossRef]
4. Contino, R.; Curtin, D.; Katz, A.; Mangano, M.L.; Panico, G.; Ramsey-Musolf, M.J.; Zanderighi, G.; Anastasiou, C.; Astill, W.; Bambhaniya, G.; et al. Physics at a 100 TeV pp collider: Higgs and EW symmetry breaking studies. *CERN Yellow Rep.* **2017**, 255–440. [CrossRef]
5. Golling, T.; Hance, M.; Harris, P.; Mangano, M.L.; McCullough, M.; Moortgat, F.; Schwaller, P.; Torre, R.; Agrawal, P.; Alves, D.S.M.; et al. Physics at a 100 TeV pp collider: beyond the Standard Model phenomena. *CERN Yellow Rep.* **2017**, 441–634. [CrossRef]
6. Fukuyama, T.; Sugiyama, H.; Tsumura, K. Constraints from muon g-2 and LFV processes in the Higgs Triplet Model. *J. High Energy Phys.* **2010**, *03*, 044. [CrossRef]
7. Mohapatra, R.N.; Pati, J.C. A Natural left–right Symmetry. *Phys. Rev.* **1975**, *D11*, 2558. [CrossRef]
8. Deshpande, N.G.; Gunion, J.F.; Kayser, B.; Olness, F.I. left–right symmetric electroweak models with triplet Higgs. *Phys. Rev.* **1991**, *D44*, 837–858. [CrossRef]
9. Duka, P.; Gluza, J.; Zralek, M. Quantization and renormalization of the manifest left–right symmetric model of electroweak interactions. *Ann. Phys.* **2000**, *280*, 336–408. [CrossRef]
10. Chakrabortty, J.; Ghosh, P.; Mondal, S.; Srivastava, T. Reconciling (g-2)$_\mu$ and charged lepton flavor violating processes through a doubly charged scalar. *Phys. Rev.* **2016**, *D93*, 115004. [CrossRef]
11. Baldini, A.M.; Bao, Y.; Baracchini, E.; Bemporad, C.; Berg, F.; Biasotti, M.; Boca, G.; Cascella, M.; Cattaneo, P.W.; Cavoto, G.; et al. Search for the lepton flavor violating decay $\mu^+ \to e^+ \gamma$ with the full dataset of the MEG experiment. *Eur. Phys. J.* **2016**, *C76*, 434. [CrossRef]
12. Aubert, B.; Karyotakis, Y.; Lees, J.P.; Poireau, V.; Prencipe, E.; Prudent, X.; Tisser, V.; Tico, J.G.; Grauges, E.; Martinelli, M.; et al. Searches for Lepton Flavor Violation in the Decays $\tau^\pm \to e^\pm \gamma$ and $\tau^\pm \to \mu^\pm \gamma$. *Phys. Rev. Lett.* **2010**, *104*, 021802. [CrossRef]
13. Bellgardt, U.; Otter, G.; Eichler, R.; Felawka, L.; Niebuhr, C.; Walter, H.K.; Bertl, W.; Lordong, N.; Martino, J.; Egli, S.; et al. Search for the Decay $\mu^+ \to e^+ e^+ e^-$. *Nucl. Phys.* **1988**, *B299*, 1–6. [CrossRef]
14. Hayasaka, K.; Inami, K.; Miyazaki, Y.; Arinstein, K.; Aulchenko, V.; Aushev, T.; Bakich, A.M.; Bay, A.; Belous, K.; Bhardwaj, V.; et al. Search for Lepton Flavor Violating Tau Decays into Three Leptons with 719 Million Produced Tau+Tau- Pairs. *Phys. Lett.* **2010**, *B687*, 139–143. [CrossRef]
15. Bertl, W.H.; Engfer, R.; Hermes, E.A.; Kurz, G.; Kozlowski, T.; Kuth, J.; Otter, G.; Rosenbaum, F.; Ryskulov, N.M.; Van Der Schaaf, A.; et al. A Search for muon to electron conversion in muonic gold. *Eur. Phys. J.* **2006**, *C47*, 337–346. [CrossRef]

16. Dinh, D.N.; Ibarra, A.; Molinaro, E.; Petcov, S.T. The $\mu - e$ Conversion in Nuclei, $\mu \to e\gamma$, $\mu \to 3e$ Decays and TeV Scale See-Saw Scenarios of Neutrino Mass Generation. *J. High Energy Phys.* **2012**, *8*, 125. [CrossRef]
17. Hisano, J.; Moroi, T.; Tobe, K.; Yamaguchi, M. Lepton flavor violation via right-handed neutrino Yukawa couplings in supersymmetric standard model. *Phys. Rev.* **1996**, *D53*, 2442–2459. [CrossRef]
18. Kitano, R.; Koike, M.; Okada, Y. Detailed calculation of lepton flavor violating muon electron conversion rate for various nuclei. *Phys. Rev.* **2002**, *D66*, 096002. [CrossRef]
19. Moore, S.R.; Whisnant, K.; Young, B.L. Second Order Corrections to the Muon Anomalous Magnetic Moment in Alternative Electroweak Models. *Phys. Rev.* **1985**, *D31*, 105. [CrossRef]
20. Leveille, J.P. The Second Order Weak Correction to (G-2) of the Muon in Arbitrary Gauge Models. *Nucl. Phys.* **1978**, *B137*, 63–76. [CrossRef]
21. Aaboud, M.; Aad, G.; Abbott, B.; Abdinov, O.; Abeloos, B.; Abidi, S.H.; AbouZeid, O.S.; Abraham, N.L.; Abramowicz, H.; Abreu, H.; et al. Search for doubly charged Higgs boson production in multi-lepton final states with the ATLAS detector using proton-proton collisions at $\sqrt{s}$ = 13 TeV. *arXiv* **2017**, arXiv:1710.09748.
22. Garayoa, J.; Schwetz, T. Neutrino mass hierarchy and Majorana CP phases within the Higgs triplet model at the LHC. *J. High Energy Phys.* **2008**, *3*, 9. [CrossRef]
23. Esteban, I.; Gonzalez-Garcia, M.; Maltoni, M.; Martinez-Soler, I.; Schwetz, T. Updated fit to three neutrino mixing: exploring the accelerator-reactor complementarity. *J. High Energy Phys.* **2017**, *1*, 87. [CrossRef]
24. NuFIT. 2017. Available online: http://www.nu-fit.org/ (accessed on 10 January 2018).
25. Sirunyan, A.M.; Tumasyan, A.; Adam, W.; Ambrogi, F.; Asilar, E.; Bergauer, T.; Brstetter, J.; Brondolin, E.; Dragicevic, M.; Ero, J.; et al. Search for a heavy right-handed W boson and a heavy neutrino in events with two same-flavor leptons and two jets at $\sqrt{s}$ = 13 TeV. *J. High Energy Phys.* **2018**, *5*, 148. [CrossRef]
26. Sirunyan, A.M.; Tumasyan, A.; Adam, W.; Ambrogi, F.; Asilar, E.; Bergauer, T.; Brstetter, J.; Dragicevic, M.; Erö, J.; Del Valle, A.E.; et al. Search for heavy neutrinos and third-generation leptoquarks in hadronic states of two τ leptons and two jets in proton-proton collisions at $\sqrt{s} = 13$ TeV. *J. High Energy Phys.* **2019**, *3*, 170. [CrossRef]
27. Das, D.; Santamaria, A. Updated scalar sector constraints in Higgs triplet model. *Phys. Rev.* **2016**, *D94*, 015015. [CrossRef]
28. Bonilla, C.; Fonseca, R.M.; Valle, J.W.F. Consistency of the triplet seesaw model revisited. *Phys. Rev.* **2015**, *D92*, 075028. [CrossRef]
29. Arhrib, A.; Benbrik, R.; Chabab, M.; Moultaka, G.; Peyranere, M.C.; Rahili, L.; Ramadan, J. The Higgs Potential in the Type II Seesaw Model. *Phys. Rev.* **2011**, *D84*, 095005. [CrossRef]
30. Krauss, M.E.; Staub, F. Unitarity constraints in triplet extensions beyond the large s limit. *arXiv* **2018**, arXiv:1805.07309.
31. Lavoura, L.; Li, L.F. Making the small oblique parameters large. *Phys. Rev.* **1994**, *D49*, 1409–1416. [CrossRef]
32. Chun, E.J.; Lee, H.M.; Sharma, P. Vacuum Stability, Perturbativity, EWPD and Higgs-to-diphoton rate in Type II Seesaw Models. *J. High Energy Phys.* **2012**, *11*, 106. [CrossRef]
33. Akeroyd, A.G.; Moretti, S. Enhancement of H to gamma gamma from doubly charged scalars in the Higgs Triplet Model. *Phys. Rev.* **2012**, *D86*, 035015. [CrossRef]
34. Shen, J.F.; Bi, Y.P.; Li, Z.X. Pair production of scalars at the ILC in the Higgs triplet model under the non-degenerate case. *EPL Europhys. Lett.* **2015**, *112*, 31002. [CrossRef]
35. Alwall, J.; Frederix, R.; Frixione, S.; Hirschi, V.; Maltoni, F.; Mattelaer, O.; Shao, H.S.; Stelzer, T.; Torrielli, P.; Zaro, M. The automated computation of tree-level and next-to-leading order differential cross-sections, and their matching to parton shower simulations. *J. High Energy Phys.* **2014**, *7*, 79. [CrossRef]
36. Rizzo, T.G. Tests of the fermion and Higgs multiplet structure of the SU(2)×U(1) model. *Phys. Rev. D* **1980**, *21*, 1404–1409. [CrossRef]
37. Czakon, M.; Gluza, J.; Jegerlehner, F.; Zralek, M. Confronting electroweak precision measurements with new physics models. *Eur. Phys. J.* **2000**, *C13*, 275–281. [CrossRef]
38. Gunion, J.F.; Grifols, J.; Mendez, A.; Kayser, B.; Olness, F.I. Higgs Bosons in left–right Symmetric Models. *Phys. Rev.* **1989**, *D40*, 1546. [CrossRef]
39. Ade, P.A.R.; Aghanim, N.; Armitage-Caplan, C.; Arnaud, M.; Ashdown, M.; Atrio-Barela, F.; Aumont, J.; Baccigalupi, C.; Banday, A.J.; Barreiro, R.B.; et al. Planck 2013 results. XVI. Cosmological parameters. *Astron. Astrophys.* **2014**, *571*, A16. [CrossRef]

40. Fileviez Perez, P.; Han, T.; Huang, G.Y.; Li, T.; Wang, K. Neutrino Masses and the CERN LHC: Testing Type II Seesaw. *Phys. Rev.* **2008**, *D78*, 015018. [CrossRef]
41. Tanabashi, M.; Hagiwara, K.; Hikasa, K.; Nakamura, K.; Sumino, Y.; Takahashi, F.; Tanaka, J.; Agashe, K.; Aielli, G.; Amsler, C.; et al. Review of Particle Physics. *Phys. Rev.* **2018**, *D98*, 030001. [CrossRef]
42. Czakon, M.; Gluza, J.; Zralek, M. Low-energy physics and left–right symmetry: Bounds on the model parameters. *Phys. Lett.* **1999**, *B458*, 355–360. [CrossRef]
43. CLIC—Compact Linear International Collider Project, CERN. Available online: http://clic-study.web.cern.ch/ (accessed on 1 December 2019).
44. Adli, E. Plasma Wakefield Linear Colliders—Opportunities and Challenges. *Philos. Trans. R. Soc.* **2019**, *337*. [CrossRef] [PubMed]
45. Nomura, T.; Okada, H.; Yokoya, H. Discriminating leptonic Yukawa interactions with doubly charged scalar at the ILC. *Nucl. Phys.* **2018**, *B929*, 193–206. [CrossRef]
46. Dev, P.S.B.; Ramsey-Musolf, M.J.; Zhang, Y. Doubly-Charged Scalars in the Type-II Seesaw Mechanism: Fundamental Symmetry Tests and High-Energy Searches. *Phys. Rev.* **2018**, *D98*, 055013. [CrossRef]
47. Bhupal Dev, P.S.; Zhang, Y. Displaced vertex signatures of doubly charged scalars in the type-II seesaw and its left–right extensions. *J. High Energy Phys.* **2018**, *10*, 199. [CrossRef]
48. Antusch, S.; Fischer, O.; Hammad, A.; Scherb, C. Low scale type II seesaw: Present constraints and prospects for displaced vertex searches. *J. High Energy Phys.* **2019**, *02*, 157. [CrossRef]
49. Acosta, D.; Adelman, J.; Affolder, T.; Akimoto, T.; Albrow, M.G.; Ambrose, D.; Amerio, S.; Amidei, D.; Anastassov, A.; Anikeev, K.; et al. Search for long-lived doubly-charged Higgs bosons in $p\bar{p}$ collisions at $\sqrt{s}$ = 1.96 TeV. *Phys. Rev. Lett.* **2005**, *95*, 071801. [CrossRef]
50. Khachatryan, V.; Sirunyan, A.M.; Tumasyan, A.; Adam, W.; Asilar, E.; Bergauer, T.; Brstetter, J.; Brondolin, E.; Dragicevic, M.; Erö, J.; et al. Search for long-lived charged particles in proton-proton collisions at $\sqrt{s}$ = 13 TeV. *Phys. Rev.* **2016**, *D94*, 112004. [CrossRef]
51. Aad, G.; Abbott, B.; Abdallah, J.; Abdinov, O.; Aben, R.; Abolins, M.; AbouZeid, O.S.; Abramowicz, H.; Abreu, H.; Abreu, R.; et al. Search for heavy long-lived multi-charged particles in pp collisions at $\sqrt{s}$ = 8 TeV using the ATLAS detector. *Eur. Phys. J.* **2015**, *C75*, 362. [CrossRef]
52. Aad, G.; Abajyan, T.; Abbott, B.; Abdallah, J.; Khalek, S.A.; Abdelalim, A.A.; Abdinov, O.; Aben, R.; Abi, B.; Abolins, M.; et al. Search for long-lived, multi-charged particles in pp collisions at $\sqrt{s}$ = 7 TeV using the ATLAS detector. *Phys. Lett.* **2013**, *B722*, 305–323. [CrossRef]
53. Abada, A.; Abbrescia, M.; AbdusSalam, S.S.; Abdyukhanov, I.; Fernez, J.A.; Abramov, A.; Aburaia, M.; Acar, A.O.; Adzic, P.R.; Agrawal, P.; et al. FCC-hh: The Hadron Collider. *Eur. Phys. J. ST* **2019**, *228*, 755–1107. [CrossRef]
54. Pospelov, M.E. FCNC in left–right symmetric theories and constraints on the right-handed scale. *Phys. Rev.* **1997**, *D56*, 259–264. [CrossRef]
55. Chakrabortty, J.; Gluza, J.; Jelinski, T.; Srivastava, T. Theoretical constraints on masses of heavy particles in left–right Symmetric Models. *Phys. Lett.* **2016**, *B759*, 361–368. [CrossRef]

© 2020 by the authors. Licensee MDPI, Basel, Switzerland. This article is an open access article distributed under the terms and conditions of the Creative Commons Attribution (CC BY) license (http://creativecommons.org/licenses/by/4.0/).

Article
Super-Weak Force and Neutrino Masses

Zoltán Trócsányi [1,2]

[1] Institute for Theoretical Physics, ELTE Eötvös Loránd University, Pázmány Péter Sétány 1/A, H-1117 Budapest, Hungary; Zoltan.Trocsanyi@cern.ch
[2] MTA-DE Particle Physics Research Group, P.O. Box 105, H-4010 Debrecen, Hungary

Received: 31 October 2019; Accepted: 2 January 2020; Published: 6 January 2020

Abstract: We consider an anomaly free extension of the standard model gauge group G_{SM} by an abelian group to $G_{SM} \otimes U(1)_Z$. The condition of anomaly cancellation is known to fix the Z-charges of the particles, but two. We fix one remaining charge by allowing for all possible Yukawa interactions of the known left-handed neutrinos and new right-handed ones that obtain their masses through interaction with a new scalar field with spontaneously broken vacuum. We discuss some of the possible consequences of the model.

Keywords: gauge symmetry; extension of the standard model of particle interactions; neutrino masses

1. Introduction

The remarkable experimental success of the standard model of elementary particle interactions [1] leaves very little room for the explanation of the observed deviations from it. This success story has culminated in the discovery of the Higgs particle [2,3], which could not have happened without the immense theoretical input to the design of the accelerator and the experiments. With this discovery, a new era of particle physics has also arrived as there is no established model that can guide us to new discoveries. Therefore, theories that might incorporate the existing deviations from the standard model are desirable.

The most outstanding experimental observations that cannot be explained by the standard model are the (i) abundance of dark matter in the universe; (ii) non-vanishing neutrino masses; (iii) leptogenesis (Baryogenesis can be explained in the standard model provided leptogenesis occurs, which is called lepto-baryogenesis); (iv) accelerating expansion of the universe, signaling the existence of dark energy [4] (There are numerous other deviations of experimental results from precision predictions, but to date none has reached the significance of discovery). In addition to (i)–(iv), (v) inflation in the early universe is also considered a fairly established fact, although there is no direct proof for it. All these facts have to be explained by such an extension of the standard model that respects (a) the high precision confirmation of the standard model at collider experiments (b) and the lack of finding new particles beyond the Higgs boson by the LHC experiments [5,6]. There is one more feature of the standard model, the metastability of a vacuum [7,8] that does not necessarily require new physics, but, if new physics exist, it should not worsen the stability, but possibly push the vacuum to the stability region.

In addition to the experimental success of the standard model, it is also highly efficient being based on the concepts of local gauge invariance and spontaneous symmetry breaking [9,10]. The only exception of economical description is the relatively large number of Yukawa couplings of the fermions needed to explain their masses. The generation of the fermion masses, however, is also highly efficient in the sense that it uses the same spontaneous symmetry breaking of the scalar field to which all other particles owe their masses. In this spirit, it is reasonable to expect that the non-vanishing masses of the neutrinos should be explained by Yukawa couplings too. In addition, the choice of the gauge groups and number of family replications look arbitrary and presently these are determined by phenomenology only.

Clearly, the neutrino masses must play a fundamental role in the possible extensions of the standard model. As the gauge and mass eigenstates of the neutrinos differ, they must feel a second force to the gauge interaction. The second force can be a Yukawa coupling to a scalar. Such explanation of neutrino masses in general requires the assumption of the existence of right-handed neutrinos and perhaps a new scalar field.

In the spirit of economy and level of arbitrariness explained above, in this article, we propose an extension of the zoo of particles in the standard model with three right-handed neutrinos and the gauge symmetry of the standard model Lagrangian $G_{SM} = SU(3)_c \otimes SU(2)_L \otimes U(1)_Y$ to $G_{SM} \otimes U(1)_Z$. Such extensions have already been considered in the literature extensively (for an incomplete set of popular examples and their studies, see [11–13]). In particular, it was shown that the charge assignment of the matter fields is constrained by the requirement of anomaly cancellations up to two free charges [14]. To define the model completely, one has to take a specific choice for these remaining free charges. In this article, we propose that the mechanism for the generation of neutrino masses fixes the values of the $U(1)_Z$ charges up to an overall scale that can be embedded in the $U(1)_Z$ coupling.

The difference between our proposal and existing studies is two-fold. The model proposed here introduces a new force along the same principles as the known forces are included in the standard model: all renormalizable terms that are allowed by the underlying gauge symmetry are present, but no other symmetry than the extra $U(1)_Z$ is assumed. Our primary goal is not the prediction of new observable phenomena at collider experiments, but first focus only on the unexplained phenomena (i–iv), with respecting the observations (a) and (b). As the deviations from the standard model are related to the intensity and cosmic frontiers of particle physics, we assume that the new $U(1)_Z$ interaction is secluded from the standard model by a small coupling. Thus, we propose the model in a region of the parameter space that has received little attention before.

2. Definition of the Model

2.1. Fermion Sector

We consider the usual three fermion families of the standard model extended with one right-handed Dirac neutrino in each family (We find it natural to assume one extra neutrino in each family although known observations do not exclude other possibilities). We introduce the notation

$$\psi_{q,1}^f = \begin{pmatrix} U^f \\ D^f \end{pmatrix}_L, \quad \psi_{q,2}^f = U_R^f \quad \psi_{q,3}^f = D_R^f; \quad \psi_{l,1}^f = \begin{pmatrix} \nu^f \\ \ell^f \end{pmatrix}_L, \quad \psi_{l,2}^f = \nu_R^f \quad \psi_{l,3}^f = \ell_R^f \quad (1)$$

for the chiral quark fields ψ_q and chiral lepton fields ψ_l. In Equation (1), L and R denote the left and right-handed projections of the same field (The Weyl spinors of ν_L and ν_R can be embedded into different Dirac spinors, leading to Majorana neutrinos, without essential changes in the model. However, the negative results of the experiments searching for neutrinoless double β-decay make the Majorana nature of neutrinos increasingly unlikely),

$$\psi_{L/R} \equiv \psi_\mp = \frac{1}{2}(1 \mp \gamma_5)\psi \equiv P_{L/R}\psi. \quad (2)$$

Then, the field content in family f ($f = 1, 2$ or 3) consists of two quarks, U_f, D_f, a neutrino ν_f and a charged lepton ℓ_f, where U_f is the generic notation for the u-type quarks $U_1 = u$, $U_2 = c$, $U_3 = t$, while D_f is that for d-type quarks, $D_1 = d$, $D_2 = s$, and $D_3 = b$. The charged leptons ℓ_f can be $\ell_1 = e$, $\ell_2 = \mu$ or $\ell_3 = \tau$ and ν_f are the corresponding neutrinos, $\nu_1 = \nu_e$, $\nu_2 = \nu_\mu$, $\nu_3 = \nu_\tau$.

For a matrix $U \in G_{SM} \otimes U(1)_Z$, the three generic fields in Equation (1) transform as

$$U\psi_1(x) = e^{iT\cdot\alpha(x)} e^{iy_1\beta(x)} e^{iz_1\zeta(x)} \psi_1(x), \quad \text{where} \quad T = \frac{1}{2}(\tau_1, \tau_2, \tau_3),$$
$$U\psi_j(x) = \phantom{e^{iT\cdot\alpha(x)}} e^{iy_j\beta(x)} e^{iz_j\zeta(x)} \psi_j(x), \quad \text{where} \quad j = 2, 3, \tag{3}$$

and $\alpha = (\alpha_1, \alpha_2, \alpha_3)$, with $\alpha_i, \beta, \zeta \in \mathbb{R}$. The matrices τ_i are the Pauli matrices, y_j is the hypercharge, while z_j denotes the Z-charge of the field ψ_j. There is a lot of freedom how to choose the Z-charges. In this article, we make two assumptions that fix these completely. The first is that the charges do not depend on the families, which is also the case in the standard model (Several recent observations hint at violation of lepton flavor universality, which may be taken into account in our model by choosing family dependent Z-charges. However, those results are controversial at present, so we neglect them). With this assumption, the assignment for the Z-charges of the fermions can be expressed using two free numbers Z_1 and Z_2 of the U quark fields if we want a model free of gauge and gravity anomalies. The rest of the charges must take values as given in Table 1 [14].

Table 1. Assignments for the representations (for $SU(N)$) and charges (for $U(1)$) of fermion and scalar fields of the complete model. The charges y_j denote the eigenvalue of $Y/2$, with Y being the hypercharge operator and z_j denote the supercharges of the fields ψ_j of Equation (1) ($j = 1, 2, 3$). The right-handed Dirac neutrinos ν_R are sterile under the G_{SM} group. The sixth column gives a particular realization of the $U(1)_Z$ charges, motivated below, and the last one is added for later convenience.

field	$SU(3)_c$	$SU(2)_L$	y_j	z_j	z_j	$r_j = z_j/z_\phi - y_j$
U_L, D_L	3	2	$\frac{1}{6}$	Z_1	$\frac{1}{6}$	0
U_R	3	1	$\frac{2}{3}$	Z_2	$\frac{7}{6}$	$\frac{1}{2}$
D_R	3	1	$-\frac{1}{3}$	$2Z_1 - Z_2$	$-\frac{5}{6}$	$-\frac{1}{2}$
ν_L, ℓ_L	1	2	$-\frac{1}{2}$	$-3Z_1$	$-\frac{1}{2}$	0
ν_R	1	1	0	$Z_2 - 4Z_1$	$\frac{1}{2}$	$\frac{1}{2}$
ℓ_R	1	1	-1	$-2Z_1 - Z_2$	$-\frac{3}{2}$	$-\frac{1}{2}$
ϕ	1	2	$\frac{1}{2}$	z_ϕ	1	$\frac{1}{2}$
χ	1	1	0	z_χ	-1	-1

The Dirac Lagrangian summed over the family replications,

$$\mathcal{L}_D = i \sum_{f=1}^{3} \sum_{j=1}^{3} \left(\overline{\psi}_{q,j}^f(x) \slashed{D}_j \psi_{q,j}^f(x) + \overline{\psi}_{l,j}^f(x) \slashed{D}_j \psi_{l,j}^f(x) \right), \tag{4}$$

$$D_j^\mu = \partial^\mu + ig_L T \cdot W^\mu + ig_Y y_j B^\mu + ig_Z z_j Z^\mu$$

is invariant under local $G = G_{SM} \otimes U(1)_Z$ gauge transformations, provided the five gauge fields introduced in the covariant derivative transform as

$$T \cdot W^\mu(x) \xrightarrow{G} T \cdot W'^\mu(x) = U(x) T \cdot W^\mu(x) U^\dagger(x) + \frac{i}{g_L}[\partial^\mu U(x)] U^\dagger(x)$$
$$B^\mu \xrightarrow{G} B'^\mu(x) = B^\mu(x) - \frac{1}{g_Y}\partial^\mu \beta(x) \quad Z^\mu \xrightarrow{G} Z'^\mu(x) = Z^\mu(x) - \frac{1}{g_Z}\partial^\mu \zeta(x), \tag{5}$$

where $U(x) = \exp[iT \cdot \alpha(x)]$. The gauge invariant kinetic term for these vector fields is

$$\mathcal{L}_{B,Z,W} = -\frac{1}{4}B_{\mu\nu}B^{\mu\nu} - \frac{1}{4}Z_{\mu\nu}Z^{\mu\nu} - \frac{1}{4}W_{\mu\nu} \cdot W^{\mu\nu}, \tag{6}$$

with $B_{\mu\nu} = \partial_\mu B_\nu - \partial_\nu B_\mu \equiv \partial_{[\mu} B_{\nu]}$, $Z_{\mu\nu} = \partial_{[\mu} Z_{\nu]}$ and $W_{\mu\nu} = \partial_{[\mu} W_{\nu]} - g\, W_\mu \times W_\nu$. The field strength $T \cdot W_{\mu\nu}$ transforms covariantly under G transformations, $T \cdot W_{\mu\nu} \xrightarrow{G} U(x)\, T \cdot W_{\mu\nu}\, U^\dagger(x)$, but $B_{\mu\nu}$ and $Z_{\mu\nu}$ are invariant, hence a kinetic mixing term of the $U(1)$ fields is also allowed by gauge invariance:

$$-\frac{\epsilon}{2} B_{\mu\nu} Z^{\mu\nu}. \qquad (7)$$

We can get rid of this mixing term by redefining the $U(1)$ fields using the transformation

$$\begin{pmatrix} B'_\mu \\ Z'_\mu \end{pmatrix} = \begin{pmatrix} 1 & \sin\theta_Z \\ 0 & \cos\theta_Z \end{pmatrix} \begin{pmatrix} B_\mu \\ Z_\mu \end{pmatrix} \qquad \sin\theta_Z = \epsilon. \qquad (8)$$

In terms of the redefined fields, the covariant derivative becomes

$$D^\mu_j = \partial^\mu + i g_L\, T \cdot W^\mu + i g_Y\, y_j B'^\mu + i(g'_Z z_j - g'_Y y_j) Z'^\mu, \qquad (9)$$

where $g'_Y = g_Y \tan\theta_Z = \epsilon g_Y + O(\epsilon^3)$ and $g'_Z = g_Z/\cos\theta_Z = g_Z + O(\epsilon^2)$. Thus, the effect of the kinetic mixing is to change the couplings of the matter fields to the vector field Z^μ. Note that we cannot immediately combine the coupling factor $(g'_Z z_j - g'_Y y_j)$ into a single product of a coupling and a charge. We shall discuss this issue further below.

Gauge symmetry forbids mass terms for gauge bosons. Fermion masses must also be absent because

$$m\,\bar\psi\psi = m\,\bar\psi_L \psi_R + m\,\bar\psi_R \psi_L,$$

but the ψ_L, ψ_R fields transform differently under G. Thus, the G-invariant Lagrangian describes massless fields in contradiction to observation.

2.2. Scalar Sector

To solve the puzzle of missing masses, we proceed similarly as in the standard model, but, in addition to the usual Brout–Englert–Higgs (BEH) field ϕ, which is an $SU(2)_L$-doublet

$$\phi = \begin{pmatrix} \phi^+ \\ \phi^0 \end{pmatrix} = \frac{1}{\sqrt{2}} \begin{pmatrix} \phi_1 + i\phi_2 \\ \phi_3 + i\phi_4 \end{pmatrix}. \qquad (10)$$

We also introduce another complex scalar χ that transforms as a singlet under G_{SM} transformations. The gauge invariant Lagrangian of the scalar fields is

$$\mathcal{L}_{\phi,\chi} = [D_{\phi\,\mu}\phi]^* D^\mu_\phi \phi + [D_{\chi\,\mu}\chi]^* D^\mu_\chi \chi - V(\phi,\chi), \qquad (11)$$

where the covariant derivative for the scalar s ($s = \phi, \chi$) is

$$D^\mu_s = \partial^\mu + i g_L\, T \cdot W^\mu + i g_Y\, y_s B'^\mu + i(g'_Z z_s - g'_Y y_s) Z'^\mu \qquad (12)$$

and the potential energy

$$V(\phi,\chi) = V_0 - \mu^2_\phi |\phi|^2 - \mu^2_\chi |\chi|^2 + \left(|\phi|^2, |\chi|^2\right) \begin{pmatrix} \lambda_\phi & \frac{\lambda}{2} \\ \frac{\lambda}{2} & \lambda_\chi \end{pmatrix} \begin{pmatrix} |\phi|^2 \\ |\chi|^2 \end{pmatrix}, \qquad (13)$$

in addition to the usual quartic terms, introduces a coupling term $-\lambda |\phi|^2 |\chi|^2$ of the scalar fields in the Lagrangian. For the doublet, $|\phi|$ denotes the length $\sqrt{|\phi^+|^2 + |\phi^0|^2}$. The value of the additive constant V_0 is irrelevant for particle dynamics but may be relevant for inflationary scenarios, hence we allow for its non-vanishing value. In order for this potential energy to be bounded from below, we have to require the positivity of the self-couplings, $\lambda_\phi, \lambda_\chi > 0$. The eigenvalues of the coupling matrix are

$$\lambda_\pm = \frac{1}{2}\left(\lambda_\phi + \lambda_\chi \pm \sqrt{(\lambda_\phi - \lambda_\chi)^2 + \lambda^2}\right), \tag{14}$$

while the corresponding un-normalized eigenvectors are

$$u^{(+)} = \begin{pmatrix} \frac{2}{\lambda}(\lambda_+ - \lambda_\chi) \\ 1 \end{pmatrix} \quad \text{and} \quad u^{(-)} = \begin{pmatrix} \frac{2}{\lambda}(\lambda_- - \lambda_\chi) \\ 1 \end{pmatrix}. \tag{15}$$

As $\lambda_+ > 0$ and $\lambda_- < \lambda_+$, in the physical region, the potential can be unbounded from below only if $\lambda_- < 0$ and $u^{(-)}$ points into the first quadrant, which may occur only when $\lambda < 0$. In this case, to ensure that the potential is bounded from below, one also has to require that the coupling matrix be positive definite, which translates into the condition

$$4\lambda_\phi \lambda_\chi - \lambda^2 > 0. \tag{16}$$

With these conditions satisfied, we can find the minimum of the potential energy at field values $\phi = v/\sqrt{2}$ and $\chi = w/\sqrt{2}$ where the vacuum expectation values (VEVs) are

$$v = \sqrt{2}\sqrt{\frac{2\lambda_\chi \mu_\phi^2 - \lambda \mu_\chi^2}{4\lambda_\phi \lambda_\chi - \lambda^2}} \qquad w = \sqrt{2}\sqrt{\frac{2\lambda_\phi \mu_\chi^2 - \lambda \mu_\phi^2}{4\lambda_\phi \lambda_\chi - \lambda^2}}. \tag{17}$$

Using the VEVs, we can express the quadratic couplings as

$$\mu_\phi^2 = \lambda_\phi v^2 + \frac{\lambda}{2}w^2 \qquad \mu_\chi^2 = \lambda_\chi w^2 + \frac{\lambda}{2}v^2 \tag{18}$$

so those are both positive if $\lambda > 0$. If $\lambda < 0$, the constraint (16) ensures that the denominators of the VEVs in Equation (17) are positive, so the VEVs have non-vanishing real values only if

$$2\lambda_\chi \mu_\phi^2 - \lambda \mu_\chi^2 > 0 \quad \text{and} \quad 2\lambda_\phi \mu_\chi^2 - \lambda \mu_\phi^2 > 0 \tag{19}$$

simultaneously, which can be satisfied if at most one of the quadratic couplings is smaller than zero. We summarize the possible cases for the signs of the couplings in Table 2.

Table 2. Possible signs of the couplings in the scalar potential $V(\phi, \chi)$ in order to have two non-vanishing real VEVs. Θ is the step function, $\Theta(x) = 1$ if $x > 0$ and 0 if $x < 0$.

$\Theta(\lambda)$	$\Theta(\lambda_\phi)$	$\Theta(\lambda_\chi)$	$\Theta(4\lambda_\phi\lambda_\chi - \lambda^2)$	$\Theta(\mu_\phi^2)$	$\Theta(\mu_\chi^2)$	$\Theta(2\lambda_\chi\mu_\phi^2 - \lambda\mu_\chi^2)\Theta(2\lambda_\phi\mu_\chi^2 - \lambda\mu_\phi^2)$
1	1	1	unconstrained	1	1	unconstrained
0	1	1	1	1	1	unconstrained
0	1	1	1	1	0	1

After spontaneous symmetry breaking of $G \to SU(3)_c \otimes U(1)_Q$, we use the following convenient parametrization for the scalar fields:

$$\phi = \frac{1}{\sqrt{2}} e^{iT \cdot \xi(x)/v} \begin{pmatrix} 0 \\ v + h'(x) \end{pmatrix} \quad \text{and} \quad \chi(x) = \frac{1}{\sqrt{2}} e^{i\eta(x)/w}(w + s'(x)). \tag{20}$$

We can use the gauge invariance of the model to choose the unitary gauge when

$$\phi'(x) = \frac{1}{\sqrt{2}} \begin{pmatrix} 0 \\ v + h'(x) \end{pmatrix} \quad \text{and} \quad \chi'(x) = \frac{1}{\sqrt{2}}(w + s'(x)), \tag{21}$$

and the vector fields are transformed according to Equation (5). With this gauge choice, the scalar kinetic term contains quadratic terms of the gauge fields from which one can identify mass parameters of the massive standard model gauge bosons proportional to the vacuum expectation value v of the BEH field and also that of a massive vector boson Z'^μ proportional to w. We can diagonalize the mass matrix (quadratic terms) of the two real scalars (h' and s') by the rotation

$$\begin{pmatrix} h \\ s \end{pmatrix} = \begin{pmatrix} \cos\theta_S & -\sin\theta_S \\ \sin\theta_S & \cos\theta_S \end{pmatrix} \begin{pmatrix} h' \\ s' \end{pmatrix}, \tag{22}$$

where, for the scalar mixing angle $\theta_S \in (-\frac{\pi}{4}, \frac{\pi}{4})$, we find

$$\sin(2\theta_S) = -\frac{\lambda vw}{\sqrt{(\lambda_\phi v^2 - \lambda_\chi w^2)^2 + (\lambda vw)^2}}. \tag{23}$$

The masses of the mass eigenstates h and s are

$$M_{h/H} = \left(\lambda_\phi v^2 + \lambda_\chi w^2 \mp \sqrt{(\lambda_\phi v^2 - \lambda_\chi w^2)^2 + (\lambda vw)^2}\right)^{1/2}, \tag{24}$$

where $M_h \leq M_H$ by convention. At this point, either h or H can be the standard model Higgs boson. A more detailed analysis of this scalar sector but within a different $U(1)_Z$ model can be found in Ref. [15] and for the present model in Ref. [16].

2.3. Fermion Masses

We already discussed that explicit mass terms of fermions would break $SU(2)_L \otimes U(1)_Y$ invariance. However, we can introduce gauge-invariant fermion-scalar Yukawa interactions (We distinguish the hypercharge Y from the index referring to Yukawa terms using different type of letters)

$$\mathcal{L}_Y = -[c_D \bar{Q}_L \cdot \phi\, D_R + c_U \bar{Q}_L \cdot \tilde{\phi}\, U_R + c_\ell \bar{L}_L \cdot \phi\, \ell_R] + \text{h.c.}, \tag{25}$$

where h.c. means Hermitian conjugate terms and the parameters c_D, c_U, c_ℓ are called Yukawa couplings that are matrices in family indices and summation over the families is understood implicitly. The dot product abbreviates scalar products of $SU(2)$ doublets:

$$\bar{Q}_L \cdot \phi \equiv (\bar{U}, \bar{D})_L \begin{pmatrix} \phi^{(+)} \\ \phi^{(0)} \end{pmatrix} \quad \bar{Q}_L \cdot \tilde{\phi} \equiv (\bar{U}, \bar{D})_L \begin{pmatrix} \phi^{(0)*} \\ -\phi^{(+)*} \end{pmatrix} \tag{26}$$

and $\bar{L} \equiv (\bar{\nu}_\ell, \bar{\ell})$. The Z-charge of the BEH field is constrained by $U(1)_Z$ invariance of the Yukawa terms to $z_\phi = Z_2 - Z_1$, which works simultaneously for all three terms.

After spontaneous symmetry breaking and fixing the unitary gauge, this Yukawa Lagrangian becomes

$$\mathcal{L}_Y = -\frac{1}{\sqrt{2}}(v + h(x))\left[c_D \bar{D}_L D_R + c_U \bar{U}_L U_R + c_\ell \bar{\ell}_L \ell_R\right] + \text{h.c.} \tag{27}$$

We see that there are mass terms with mass matrices $M_i = \frac{c_i v}{\sqrt{2}}$, where $i = D, U, \ell$:

$$\mathcal{L}_Y = -\left(1 + \frac{h(x)}{v}\right)\left[\bar{D}_L M_D D_R + \bar{U}_L M_U U_R + \bar{\ell}_L M_\ell \ell_R\right] + \text{h.c.} \tag{28}$$

The general complex matrices M_i can be diagonalized employing bi-unitary transformations. The diagonal elements on the basis of mass eigenstates provide the mass parameters of the fermions. Due to the bi-unitary transformation, the left- and right-handed components of the fermion field are different linear combinations of the mass eigenstates.

The neutrino oscillation experiments suggest non-vanishing neutrino masses and the weak and mass eigenstates of the left-handed neutrinos do not coincide. In principle, the charge assignment of our model allows for the following gauge invariant Yukawa terms of dimension four operators for the neutrinos

$$\mathcal{L}_Y^\nu = -\sum_{i,j}\left((c_\nu)_{ij}\bar{L}_{i,L}\cdot\tilde{\phi}\,\nu_{j,R} + \frac{1}{2}(c_R)_{ij}\overline{\nu_{i,R}^c}\nu_{j,R}\chi\right) + \text{h.c.} \tag{29}$$

for arbitrary values of Z_1 and Z_2 if the superscript c denotes the charge conjugate of the field, $\nu^c = -i\gamma_2\nu^*$, and the Z-charge of the right-handed neutrinos and the new scalar satisfy the relation $z_\chi = -2z_{\nu_R}$. There are two natural choices to fix the Z-charges: (i) the left- and right-handed neutrinos have the same charge; or (ii) those have opposite charges (We explain in Section 2.5 the reason for considering this choice being natural). In the first case, we have

$$Z_2 - 4Z_1 = -3Z_1, \tag{30}$$

which is solved by $Z_1 = Z_2$, and it leads to the charge assignment of the $U(1)_{B-L}$ extension of the standard model, studied in detail (see for instance [17] and references therein). In the second case,

$$Z_2 - 4Z_1 = 3Z_1, \tag{31}$$

which is solved by $Z_1 = Z_2/7$. As the overall scale of the Z-charges depends only on the value of the gauge coupling g'_Z, we set Z_2 freely. For instance, choosing $Z_2 = 7/6$ implies $Z_1 = 1/6$ and the Z-charge of the BEH scalar is

$$z_\phi = 1, \tag{32}$$

while that of the new scalar is

$$z_\chi = -1 = -z_\phi. \tag{33}$$

While we cannot exclude the infinitely many cases when the magnitudes of Z-charges of the left- and right-handed neutrinos differ, we find it natural to assume that Equation (31) is valid. The corresponding Z-charges are given explicitly in the sixth column of Table 1.

After the spontaneous symmetry breaking of the vacuum of the scalar fields, Equation (29) leads to the following mass terms for the neutrinos:

$$\mathcal{L}_Y^\nu = -\frac{1}{2}\sum_{i,j}\left[\left(\overline{\nu_L},\overline{\nu_R^c}\right)_i M(h,s)_{ij} \begin{pmatrix} \nu_L^c \\ \nu_R \end{pmatrix}_j + \text{h.c.}\right], \tag{34}$$

where

$$M(h,s)_{ij} = \begin{pmatrix} 0 & m_D\left(1+\frac{h}{v}\right) \\ m_D\left(1+\frac{h}{v}\right) & M_M\left(1+\frac{s}{w}\right) \end{pmatrix}_{ij}, \tag{35}$$

with complex m_D and real M_M being symmetric 3×3 matrices, so $M(0,0)$ is a complex symmetric 6×6 matrix. The diagonal elements of the mass matrix $M(0,0)$ provide Majorana mass terms for the left-handed and right-handed neutrinos. Thus, we conclude that the model predicts *vanishing masses of the left-handed neutrinos* at the fundamental level.

The off-diagonal elements represent interaction terms that look formally like Dirac mass terms, $-\sum_{i,j} \overline{\nu_{i,L}}(m_D)_{ij}\nu_{j,R}+$ h.c. After spontaneous symmetry breaking the quantum numbers of the particles $\nu_{i,L}^c$ and $\nu_{i,R}$ being identical, they can mix. Thus, the propagating states will be a mixture of the left- and right-handed neutrinos, providing effective masses for the left-handed ones. Those states can be obtained by the diagonalization of the full matrix $M(0,0)$, for which a possible parametrization is given for instance in Ref. [18].

In order to understand the structure of the matrix $M(0,0)$ better, we first diagonalize the matrices m_D and M_M separately by a unitary transformation and an orthogonal one. Defining

$$\nu'_{L,i} = \sum_j (U_L)_{ij}\nu_{L,j} \quad \text{and} \quad \nu'_{R,i} = \sum_j (O_R)_{ij}\nu_{R,j}, \tag{36}$$

we can rewrite the neutrino Yukawa Lagrangian as

$$\mathcal{L}_Y^\nu = -\frac{1}{2}\sum_{i,j}\left[\left(\overline{\nu'_L}, \overline{\nu'^c_R}\right)_i M'(h,s)_{ij}\begin{pmatrix}\nu'^c_L\\ \nu'_R\end{pmatrix}_j + \text{h.c.}\right], \tag{37}$$

where

$$M'(h,s) = \begin{pmatrix} 0 & mV\left(1+\frac{h}{v}\right) \\ V^\dagger m\left(1+\frac{h}{v}\right) & M\left(1+\frac{s}{w}\right) \end{pmatrix}. \tag{38}$$

In Equation (38), m and M are real diagonal matrices, while $V = U_L^T O_R$ is a unitary matrix, $VV^\dagger = 1$, so $M'(0,0)$ is Hermitian with real eigenvalues that are the masses of the mass eigenstates of neutrinos. In general, $M'(0,0)$ may have 15 independent parameters: m_i and M_i ($i=1,2,3$), while there are three Euler angles and six phases V. Three phases can be absorbed into the definition of ν'_L.

Assuming the hierarchy $m_i \ll M_j$, we can integrate out the right-handed (heavy) neutrinos and obtain an effective higher dimensional operator with Majorana mass terms for the left-handed neutrinos

$$\mathcal{L}_{\text{dim}-5}^\nu = -\frac{1}{2}\sum_i m_{M,i}\left(1+\frac{h}{v}\right)^2\left(\overline{\nu'^c_{i,L}}\nu'_{i,L} + \text{h.c.}\right). \tag{39}$$

The Majorana masses $m_{M,i}$, i.e., eigenstates of the matrix $m_D^\dagger M_M^{-1} m_D$, are suppressed by the ratios m_i/M_i as compared to m_i. The latter has a similar role in the Lagrangian as the mass parameters of the charged leptons, so one may assume $m_i \sim O(100\,\text{keV})$, while the masses of the right-handed neutrinos can be naturally around $O(100\,\text{GeV})$, so that $m_i/M_i \sim O(10^{-6\pm 1})$ and $m_{M,i} \lesssim 0.1$ eV. Thus, if $m_i \ll M_i$, then the mixing between the light and heavy neutrinos will be very small, the $\nu'_{i,L}$ can be considered as the mass eigenstates that are mixtures of the left-handed weak eigenstates, and whose masses can be small naturally as suggested by phenomenological observations.

As we can only observe neutrinos together with their flavors through their charged current interactions, it is more natural to use the flavor eigenstates than the mass eigenstates. In the flavor basis, the couplings of the leptons to the W boson are diagonal:

$$\mathcal{L}_{CC}^{(\ell)} = -\frac{g_L}{\sqrt{2}}\sum_f \overline{\nu_L}^f W^\dagger \ell_L^f + \text{h.c.} \tag{40}$$

with summation over the three lepton flavors $f = e$, μ and τ. The same charged current interactions in mass basis $\nu_{L,i} = (U_{PMNS})_{if} \nu_L^f$, contain the Pontecorvo–Maki–Nakagawa–Sakata matrix U_{PMNS},

$$\mathcal{L}_{CC}^{(\ell)} = -\frac{g_L}{\sqrt{2}} \sum_{i,f=1}^{3} \overline{\nu_{L,i}} \, (U_{PMNS})_{if} \, W^{\dagger} \, \ell_L^f + \text{h.c.} \tag{41}$$

just like the charged current quark interactions contain the Cabibbo–Kobayashi–Maskawa matrix. If the heavy neutrinos are integrated out, then the matrix U_L coincides with the PMNS matrix. For propagating degrees of freedom, such as in the case of traveling neutrinos over macroscopic distances, one should use mass eigenstates $\nu_{L,i}$ and the PMNS matrix becomes the source of neutrino oscillations in flavor space. However, in the case of elementary particle scattering processes involving the left-handed neutrinos, one can work using the flavor basis, i.e., with Equation (40) because the effect of their masses can be neglected.

2.4. Re-Parametrization into Right-Handed and Mixed Couplings

Having set the Z-charges of the matter fields, we can re-parametrize the couplings to Z' using the new coupling

$$g'_{ZY} = g'_Z - g'_Y = \frac{g_Z - g_Y \sin \theta_Z}{\cos \theta_Z} = g_Z - \epsilon g_Y + O(\epsilon^2), \tag{42}$$

with ϵ being the strength of kinetic mixing. Then, the covariant derivative in Equation (9) becomes

$$D_j^\mu = \partial^\mu + i g_L \, \mathbf{T} \cdot \mathbf{W}^\mu + i y_j g_Y B'^\mu + i \left(r_j g'_Z + y_j g'_{ZY} \right) Z'^\mu, \tag{43}$$

where $r_j = z_j - y_j$ and its values are given explicitly in the last column of Table 1. Thus, if a $U(1)_Z$ extension of G_{SM} is free of gauge and gravity anomalies and the Z-charges of left and right-handed fields are the opposite, then it is equivalent to a $U(1)_R$ extension with tree-level mixed coupling g'_{ZY} [19], related to the kinetic mixing parameter ϵ by Equation (42).

Particle phenomenology of the standard model suggests that the interaction of the fermions through the Z' vector boson must be suppressed significantly. The origin of such a suppression can be either a small coupling to Z' or the large mass of Z'. Usual studies in the literature focus on the latter case. Here, we suggest to focus on the former possibility.

The complete Lagrangian is the sum of the pieces given in Equations (4), (6), (11), (25) and (29),

$$\mathcal{L} = \mathcal{L}_D + \mathcal{L}_{B,Z,W} + \mathcal{L}_{\phi,\chi} + \mathcal{L}_Y + \mathcal{L}_Y^\nu, \tag{44}$$

with covariant derivative given in Equation (43), i.e., the kinetic mixing of Equation (7) is also taken into account.

2.5. Mixing in the Neutral Gauge Sector

The neutral gauge fields of the standard model and the Z' mix, which leads to mass eigenstates A_μ, Z_μ and T_μ (not to be confused with the isospin components T_i, $i = 1, 2, 3$). The mixing is described by a 3×3 mixing matrix as

$$\begin{pmatrix} W_\mu^3 \\ B'_\mu \\ Z'_\mu \end{pmatrix} = \begin{pmatrix} \cos\theta_W \cos\theta_T & -\cos\theta_W \sin\theta_T & \sin\theta_W \\ -\sin\theta_W \cos\theta_T & \sin\theta_W \sin\theta_T & \cos\theta_W \\ \sin\theta_T & \cos\theta_T & 0 \end{pmatrix} \begin{pmatrix} Z_\mu \\ T_\mu \\ A_\mu \end{pmatrix}. \tag{45}$$

For the Weinberg mixing angle θ_W, we have the usual value $\sin\theta_W = g_Y/\sqrt{g_L^2 + g_Y^2}$. We introduce the notion of reduced coupling defined by $\gamma_i = g_i/g_L$, i.e., $\gamma_L = 1$. Then, we have

$$\sin\theta_W = \frac{\gamma_Y}{\sqrt{1+\gamma_Y^2}}, \qquad \cos\theta_W = \frac{1}{\sqrt{1+\gamma_Y^2}} \tag{46}$$

and, for the mixing angle θ_T of the Z' boson, we find

$$\sin\theta_T = \left[\frac{1}{2}\left(1 - \frac{1-\kappa^2-\tau^2}{\sqrt{(1+\kappa^2+\tau^2)^2 - 4\tau^2}}\right)\right]^{1/2},$$
$$\cos\theta_T = \left[\frac{1}{2}\left(1 + \frac{1-\kappa^2-\tau^2}{\sqrt{(1+\kappa^2+\tau^2)^2 - 4\tau^2}}\right)\right]^{1/2}, \tag{47}$$

so $\tan(2\theta_T) = 2\kappa/(1-\kappa^2-\tau^2)$, with

$$\kappa = \frac{\gamma_Y' - 2\gamma_Z'}{\sqrt{1+\gamma_Y^2}} \qquad \tau = 2\frac{\gamma_Z' \tan\beta}{\sqrt{1+\gamma_Y^2}} \tag{48}$$

and

$$\tan\beta = \frac{w}{v} \tag{49}$$

is the ratio of the scalar vacuum expectation values (not a scalar mixing angle). For small values of the new couplings γ_{ZY}' and γ_Z', implying small κ, we have

$$\theta_T = \kappa + O(\tau^2, \kappa^3). \tag{50}$$

The charged current interactions remain the same as in the standard model. The neutral current Lagrangian can be written in the form

$$\mathcal{L}_{NC} = \mathcal{L}_{QED} + \mathcal{L}_Z + \mathcal{L}_T, \tag{51}$$

where the first term is the usual Lagrangian of QED,

$$\mathcal{L}_{QED} = -eA_\mu J_{em}^\mu \qquad J_{em}^\mu = \sum_{f=1}^{3}\sum_{j=1}^{3} e_j\left(\overline{\psi}_{q,j}^f(x)\gamma^\mu \psi_{q,j}^f(x) + \overline{\psi}_{l,j}^f(x)\gamma^\mu \psi_{l,j}^f(x)\right). \tag{52}$$

The second one is a neutral current coupled to the Z^0 boson,

$$\mathcal{L}_Z = -eZ_\mu\left(\cos\theta_T J_Z^\mu + \sin\theta_T J_T^\mu\right) = -eZ_\mu J_Z^\mu + O(\theta_T), \tag{53}$$

and the third one is the neutral current coupled to the T^0 boson,

$$\mathcal{L}_T = -eT_\mu\left(-\sin\theta_T J_Z^\mu + \cos\theta_T J_T^\mu\right). \tag{54}$$

In Equation (52), e is the electric charge unit and e_j is the electric charge of field ψ_j in units of e. In Equations (53) and (54), J_Z^μ is the usual neutral current,

$$J_Z^\mu = \sum_{f=1}^{3}\sum_{j=1}^{3} \frac{T_3 - \sin^2\theta_W e_j}{\sin\theta_W \cos\theta_W}\left(\overline{\psi}_{q,j}^f(x)\gamma^\mu \psi_{q,j}^f(x) + \overline{\psi}_{l,j}^f(x)\gamma^\mu \psi_{l,j}^f(x)\right), \tag{55}$$

while the new neutral current has the same dependence on fermion dynamics with different coupling strength:

$$J_T^\mu = \sum_{f=1}^{3}\sum_{j=1}^{3} \frac{\gamma'_Z r_j + \gamma'_{ZY} y_j}{\sin\theta_W} \left(\overline{\psi}_{q,j}^f(x)\gamma^\mu \psi_{q,j}^f(x) + \overline{\psi}_{l,j}^f(x)\gamma^\mu \psi_{l,j}^f(x) \right). \tag{56}$$

We can rewrite these currents as vector–axialvector currents using the non-chiral fields ψ_f

$$J_X^\mu = \sum_f \overline{\psi}_f(x)\gamma^\mu (v_f^{(X)} - a_f^{(X)}\gamma_5)\psi_f(x) \quad X = Z \text{ or } T \tag{57}$$

with vector couplings $v_f^{(X)}$ and axialvector couplings $a_f^{(X)}$ given in Appendix A and the summation runs over all quark and lepton flavors. Clearly, the QED current J_{em}^μ can also be written using non-chiral fields in the form of Equation (57) with $v_f^{(em)} = e_f$ and $a_f^{(em)} = 0$.

As the dependence on the couplings and charges of the neutral currents in Equations (55) and (56) are very different for different fermion fields, the only way that the standard model phenomenology is not violated by the extended model is if θ_T is small, which supports the expansion used in Equation (53). The choice for the Z-charges made in Equation (31) leads to the current J_T^μ being chiral, which we find natural as it mixes with the other chiral current J_Z^μ according to Equations (53) and (54).

To define the perturbation theory of this model explicitly, we present the Feynman rules in Appendix A.

2.6. Masses of the Gauge Bosons

The photon is massless, while the masses of the massive neutral bosons are

$$M_Z = M_W \frac{\cos\theta_T}{\cos\theta_W}\left[(1 - \kappa\tan\theta_T)^2 + (\tau\tan\theta_T)^2\right]^{1/2} \tag{58}$$

and

$$M_T = M_W \frac{\sin\theta_T}{\cos\theta_W}\left[(1 + \kappa\cot\theta_T)^2 + (\tau\cot\theta_T)^2\right]^{1/2}, \tag{59}$$

where $M_W = \frac{1}{2}vg_L$, and we assumed $M_T < M_Z$. Indeed, in order to have M_Z within the experimental uncertainty of the known measured value, we need $\theta_T \simeq 0$, which justifies the expansions at $\kappa = 0$,

$$M_Z = \frac{M_W}{\cos\theta_W}\left(1 + O(\kappa^2)\right) \simeq \frac{M_W}{\cos\theta_W} \tag{60}$$

and

$$M_T = \frac{M_W}{\cos\theta_W}\tau\left(1 + O(\kappa^2)\right) \simeq M_{Z'}, \tag{61}$$

where we used Equation (50) and $M_{Z'} = wg'_Z$. Thus, τ can also be written as the ratio of the masses of the two massive neutral gauge bosons,

$$\tau = \frac{M_{Z'}}{M_W}\cos\theta_W \simeq \frac{M_T}{M_Z}, \tag{62}$$

justifying our assumption on the hierarchy of masses. In fact, unless $w \gg v$, we find $M_T \ll M_Z$.

2.7. Free Parameters

There are five parameters in the scalar sector, λ_ϕ, λ_χ, λ, v and w that has to be determined experimentally, while the values of μ_ϕ and μ_χ (at tree level) are given in Equation (18). However, it is more convenient to use parameters that can be measured more directly, for instance,

$$M_h \quad M_H \quad \sin\theta_S \quad v = (\sqrt{2}G_F)^{-1/2} \text{ and } \tan\beta, \tag{63}$$

of which we know two from measurements: one of the scalar masses and Fermi's constant.

In addition to the neutrino Yukawa couplings (or neutrino masses and PMNS mixing parameters), there are five free parameters in the model that we choose as the mass of the new scalar particle M_h or M_H (the other being fixed by the mass of the Higgs boson), the scalar and vector mixing angles $\sin\theta_S$ and $\sin\theta_T$, the ratio of the vacuum expectation values $\tan\beta$ and τ that is essentially the new gauge coupling. It can be shown [16] that, requiring stable vacuum up to the Planck scale, the Higgs particle coincides with the scalar h and according to a one-loop analysis of the running scalar couplings M_H falls into the range [144,558] GeV.

The other parameters can be expressed in terms of the free ones as follows: $w = v\tan\beta$,

$$\lambda_\phi = \frac{1}{2v^2}\left(M_{h/H}^2 \cos^2\theta_S + M_{H/h}^2 \sin^2\theta_S\right),$$

$$\lambda_\chi = \frac{1}{2w^2}\left(M_{H/h}^2 \cos^2\theta_S + M_{h/H}^2 \sin^2\theta_S\right), \tag{64}$$

$$\lambda = \sin(2\theta_S)\frac{M_H^2 - M_h^2}{2vw}$$

(first indices are to be used if $\lambda_\phi v^2 < \lambda_\chi w^2$, the second ones otherwise). The new parameters in the gauge sector can be expressed as

$$\tan\theta_Z = \frac{\tau - \kappa\tan\beta}{\tan\beta\sin\theta_W} \quad \gamma'_Z = \frac{\tau}{2\tan\beta\cos\theta_W} \quad \gamma'_Y = \frac{\tau - \kappa\tan\beta}{\tan\beta\cos\theta_W} \quad \gamma'_{ZY} = \frac{2\kappa\tan\beta - \tau}{2\tan\beta\cos\theta_W}, \tag{65}$$

$$\kappa = \cot(2\theta_T)\left(\sqrt{1 + (1-\tau^2)\tan^2(2\theta_T)} - 1\right) = (1 - \tau^2)\sin\theta_T + O(\theta_T^3).$$

3. Discussion

Our hope in devising this model is to explain the established experimental observations listed in the introduction. We envisage the following scenario:

- The lightest new particle is a natural candidate for WIMP dark matter if it is sufficiently stable.
- Majorana neutrino mass terms for the right-handed neutrinos and Yukawa interactions between the left- and right-handed neutrinos and the BEH vacuum are generated by the spontaneous symmetry breaking of the scalar fields as outlined in Section 2.3. This scenario provides a possible origin of neutrino oscillations and effective Majorana mass terms for the left-handed neutrinos.
- The neutrino Yukawa terms provide a source for the PMNS matrix as shown in Section 2.3, which can have a CP-violating phase yielding stronger CP violation in the lepton sector than there is in the quark sector.
- The vacuum of the χ scalar has a charge $z_j = -1$ (or $r_j = -1$) that may be a source of the current accelerated expansion of the universe.
- The second scalar together with the established BEH field can cause hybrid inflation.

At present, we consider these possible consequences of the model that need further studies to find out if they fulfill. Before exploring that the model makes these explanations credible, we have to find answer to the following question: *Is there any region of the parameter space of the model that is not*

4. Conclusions

In this paper, we collected the well established experimental observations that cannot be explained by the standard model of particle interactions. We have then proposed an anomaly free extension by a $U(1)_Z$ gauge group, which is the simplest possible model. We also assumed the existence of a new complex scalar field with Z-charge only (i.e., neutral with respect to the standard model interactions) and three right-handed neutrinos. In order to fix the Z-charges of the particle spectrum, we assumed that the left- and right-handed neutrinos have opposite Z-charges. Thus, such a model predicts the existence of (i) a massive neutral vector boson; (ii) a massive scalar particle and (iii) three massive right-handed neutrinos. The left-handed neutrinos remain massless as in the standard model, but their Yukawa interactions with the BEH field and the right-handed neutrinos provide a field theoretical basis for explaining neutrino oscillations and predict effective Majorana masses for the propagating mass eigenstates.

We have discussed how the new neutral gauge field Z^μ mixes with those of the standard model (B^μ and W_3^μ) and argued that the mixing results in a new vector boson T^0 of a small mass related to the small new gauge coupling and small mixing with the standard model vector fields. We also presented the Feynman rules of the model in unitary gauge and collected the new free parameters.

In order for the predictions of the model be credible, we have to answer whether there is any region of the parameter space that is not excluded by experimental results established in standard model phenomenology or elsewhere. To answer such a question with satisfaction, studies well beyond the scope of a single article are needed, which forecasts an exciting research project.

Funding: This work was supported by grant K 125105 of the National Research, Development and Innovation Fund in Hungary.

Acknowledgments: I am grateful to G. Cynolter, D. Horváth, S. Iwamoto, A. Kardos and S. Katz for their constructive criticism on the manuscript.

Conflicts of Interest: The author declares no conflict of interest.

Abbreviations

BEH	Brout–Englert–Higgs
PMNS	Pontecorvo–Maki–Nakagawa–Sakata
QCD	quantum chromodynamics
QED	quantum electrodynamics
SM	standard model
SSB	spontaneous symmetry breaking
UV	ultra-violet
VEV	vacuum expectation value

Appendix A. Feynman Rules

The Feynman rules of the model are obtained from the complete Lagrangian in Equation (44). For studying the UV behaviour of the model, it is convenient to use the Feynman rules before SSB, while for low energy phenomenology the rules after SSB are needed. In this paper, we present only the latter in a unitary gauge. The propagators of the new fields are related trivially to those of the standard fields. Thus, we present only the vertices, neglecting the rules related to QCD, which are unchanged.

- Gauge field–fermion interactions $V_\alpha \bar{f}_i f_j$: $-ie\gamma_\alpha(C^- P_- + C^+ P_+)$, where $C^\pm$ depends on the type of the gauge boson participating in the interaction, the flavor f of fermions and family number i and j as follows:

$V\bar{f}_i f_j$	C^+	C^-
$\gamma \bar{f}_i f_j$	$e_f \delta_{ij}$	$e_f \delta_{ij}$
$Z\bar{f}_i f_j$	$(g_f^+ \cos\theta_T + h_f^+ \sin\theta_T)\delta_{ij}$	$(g_f^- \cos\theta_T + h_f^- \sin\theta_T)\delta_{ij}$
$T\bar{f}_i f_j$	$(-g_f^+ \sin\theta_T + h_f^+ \cos\theta_T)\delta_{ij}$	$(-g_f^- \sin\theta_T + h_f^- \cos\theta_T)\delta_{ij}$
$W^+ \bar{u}_i d_j$	0	$\dfrac{1}{\sqrt{2}\sin\theta_W} V_{ij}$
$W^- \bar{d}_j u_i$	0	$\dfrac{1}{\sqrt{2}\sin\theta_W} V_{ij}^\dagger$
$W^+ \bar{\nu}_i \ell_j$	0	$\dfrac{1}{\sqrt{2}\sin\theta_W} \delta_{ij}$
$W^- \bar{\ell}_j \nu_i$	0	$\dfrac{1}{\sqrt{2}\sin\theta_W} \delta_{ij}$,

where

$$g_f^+ = -\frac{\sin\theta_W}{\cos\theta_W} e_f \quad g_f^- = \frac{T_f^3 - \sin^2\theta_W e_f}{\sin\theta_W \cos\theta_W} \quad h_f^\pm = \frac{\gamma_Z' R_f^\pm + \gamma_{ZY}'(e_f - R_f^\mp)}{\sin\theta_W}, \quad (A1)$$

where $R_f^+ = 1/2$ for U_f or ν_f, $R_f^+ = -1/2$ for D_f or ℓ_f and $R_f^- = 0$. The vector and axial vector couplings of the Z^0 boson read as

$$v_f^{(Z)} = \frac{1}{2}\left(g_f^- + g_f^+\right)\cos\theta_T + \frac{1}{2}\left(h_f^- + h_f^+\right)\sin\theta_T$$

$$= \frac{\left(T_f^3 - 2(\sin\theta_W)^2 e_f\right)\cos\theta_T + \left(\kappa e_f + \gamma_Y'(R_f^+ - e_f)\cos\theta_W\right)\sin\theta_T}{2\sin\theta_W \cos\theta_W}$$

$$= \frac{T_f^3 - 2(\sin\theta_W)^2 e_f}{2\sin\theta_W \cos\theta_W} + O(\theta_T),$$

$$a_f^{(Z)} = \frac{1}{2}\left(g_f^- - g_f^+\right)\cos\theta_T + \frac{1}{2}\left(h_f^- - h_f^+\right)\sin\theta_T = \frac{T_f^3 \cos\theta_T - \kappa R_f^+ \sin\theta_T}{2\sin\theta_W \cos\theta_W}$$

$$= \frac{T_f^3}{2\sin\theta_W \cos\theta_W} + O(\theta_T),$$

while those of the T^0 boson are

$$v_f^{(T)} = \frac{\left(\kappa e_f + \gamma_Y'(R_f^+ - e_f)\cos\theta_W\right)\cos\theta_T - \left(T_f^3 - 2(\sin\theta_W)^2 e_f\right)\sin\theta_T}{2\sin\theta_W \cos\theta_W},$$

$$a_f^{(T)} = -\frac{\kappa R_f^+ \cos\theta_T + T_f^3 \sin\theta_T}{2\sin\theta_W \cos\theta_W}. \quad (A2)$$

- $H\bar{f}_i f_j$ vertex: ieC, where

$$C = -\delta_{ij}\frac{1}{2\sin\theta_W}\frac{m_{f,i}}{M_W}.$$

- $S\overline{\nu^c}_{R,i}\nu_{R,j}$ vertex: ieC, where

$$C = -\delta_{ij}\frac{1}{2\sin\theta_W \tan\beta}\frac{m_{\nu_R,i}}{M_W}.$$

- Gauge field interactions:
  - The cubic gauge field interactions of fields $V_{1,\alpha}$, $V_{2,\beta}$ and $V_{3,\gamma}$ with all-incoming kinematics, $p^\mu + q^\mu + r^\mu = 0$ are $\Gamma_{\alpha,\beta,\gamma}(p,q,r) = ieCV_{\alpha,\beta,\gamma}(p,q,r)$, where

$$V_{\alpha,\beta,\gamma}(p,q,r) = (p-q)_\gamma g_{\alpha\beta} + (q-r)_\alpha g_{\beta\gamma} + (r-p)_\beta g_{\alpha\gamma},$$

  while C depends on the type of the gauge bosons participating in the interaction as follows:

$V_1 V_2 V_3$	C
$\gamma W^+ W^-$	1
$Z W^+ W^-$	$\dfrac{\cos\theta_W}{\sin\theta_W}\cos\theta_T$
$T W^+ W^-$	$-\dfrac{\cos\theta_W}{\sin\theta_W}\sin\theta_T$

  - The quartic gauge field interactions of fields $V_{1,\alpha}$, $V_{2,\beta}$, $V_{3,\gamma}$ and $V_{4,\delta}$ are $\Gamma_{\alpha,\beta,\gamma,\delta} = ie^2 C \left[2 g_{\alpha\beta} g_{\gamma\delta} - g_{\alpha\gamma} g_{\beta\delta} - g_{\alpha\delta} g_{\beta\gamma}\right]$, where C again depends on the type of the gauge bosons participating in the interaction as follows:

$V_1 V_2 V_3 V_4$	C
$W^+ W^- \gamma\gamma$	-1
$W^+ W^- \gamma Z$	$-\dfrac{\cos\theta_W}{\sin\theta_W}\cos\theta_T$
$W^+ W^- \gamma T$	$\dfrac{\cos\theta_W}{\sin\theta_W}\sin\theta_T$
$W^+ W^- ZZ$	$-\left(\dfrac{\cos\theta_W}{\sin\theta_W}\cos\theta_T\right)^2$
$W^+ W^- TZ$	$\left(\dfrac{\cos\theta_W}{\sin\theta_W}\right)^2\cos\theta_T \sin\theta_T$
$W^+ W^- TT$	$-\left(\dfrac{\cos\theta_W}{\sin\theta_W}\sin\theta_T\right)^2$
$W^+ W^+ W^- W^-$	$\dfrac{1}{(\sin\theta_W)^2}$

- Scalar interactions: We denote the standard model Higgs boson by $\mathcal{H}$, while the new one by $\mathcal{S}$.
  - Cubic scalar interactions can be either of the form $ie\frac{C}{3!}S^3$ where C depends on the type of the scalar boson participating in the interaction:

SSS	C
$\mathcal{H}\mathcal{H}\mathcal{H}$	$-\dfrac{3}{2}\dfrac{M_h^2 \cos^2\theta_S + M_H^2 \sin^2\theta_S}{\sin\theta_W M_W}$
$\mathcal{S}\mathcal{S}\mathcal{S}$	$-\dfrac{3}{2}\dfrac{M_h^2 \sin^2\theta_S + M_H^2 \cos^2\theta_S}{\sin\theta_W M_W \tan\beta}$

or of the form $ie\frac{C}{2!}SSS'$, where C depends on the type of the S boson participating in the interaction:

SSS'	C
$\mathcal{H}\mathcal{H}S$	$-\sin\theta_S \cos\theta_S \dfrac{M_H^2 - M_h^2}{2\sin\theta_W M_W}$
$SS\mathcal{H}$	$-\sin\theta_S \cos\theta_S \dfrac{M_H^2 - M_h^2}{2\sin\theta_W M_W \tan\beta}.$

Recall that $M_{H/h}$ is the mass of the heavier/lighter scalar.

- The quartic scalar interactions are either of the form $ie^2 \frac{C}{4!} S^4$, where C depends on the type of the scalar bosons participating in the interaction as follows:

$SSSS$	C
$\mathcal{H}\mathcal{H}\mathcal{H}\mathcal{H}$	$-\dfrac{3}{4}\dfrac{M_h^2 \cos^2\theta_S + M_H^2 \sin^2\theta_S}{(\sin\theta_W M_W)^2}$
$SSSS$	$-\dfrac{3}{4}\dfrac{M_h^2 \sin^2\theta_S + M_H^2 \cos^2\theta_S}{(\sin\theta_W M_W \tan\beta)^2}$

or of the form $ie^2 \frac{C}{2!2!}\mathcal{H}^2 S^2$, where

$$C = -\frac{3}{4}\frac{M_h^2 - M_h^2}{(\sin\theta_W M_W)^2 \tan\beta}.$$

- Mixed gauge field-scalar interactions:
  - The cubic gauge field-scalar interactions of fields $V_{1,\alpha}$, $V_{2,\beta}$ and S are $ieg_{\alpha\beta}C$, where C depends on the types of the fields participating in the interaction as follows:

$V_1 V_2 S$	C
$W^+ W^- \mathcal{H}$	$\dfrac{M_W}{\sin\theta_W}$
$ZZ\mathcal{H}$	$\dfrac{M_W}{\sin\theta_W} \dfrac{(\cos\theta_T - \kappa\sin\theta_T)^2}{(\cos\theta_W)^2}$
$TT\mathcal{H}$	$\dfrac{M_W}{\sin\theta_W} \dfrac{(\sin\theta_T + \kappa\cos\theta_T)^2}{(\cos\theta_W)^2}$
$TZ\mathcal{H}$	$\dfrac{M_W}{\sin\theta_W} \dfrac{(\sin\theta_T + \kappa\cos\theta_T)(\kappa\sin\theta_T - \cos\theta_T)}{(\cos\theta_W)^2}$
ZZS	$\dfrac{M_W}{\sin\theta_W \tan\beta} \dfrac{(\tau\sin\theta_T)^2}{(\cos\theta_W)^2}$
TTS	$\dfrac{M_W}{\sin\theta_W \tan\beta} \dfrac{(\tau\cos\theta_T)^2}{(\cos\theta_W)^2}$
TZS	$\dfrac{M_W}{\sin\theta_W} \dfrac{\tau^2 \sin\theta_T \cos\theta_T}{(\cos\theta_W)^2}.$

- Quartic gauge field-scalar interactions $V_\alpha V_\beta SS : ie^2 g_{\alpha\beta} C$, where C depends on the type of the gauge boson participating in the interaction as follows:

$V_1 V_2 SS$	C
$W^+ W^- \mathcal{HH}$	$\dfrac{1}{2(\sin\theta_W)^2}$
$ZZ\mathcal{HH}$	$\dfrac{(\cos\theta_T - \kappa \sin\theta_T)^2}{2(\cos\theta_W \sin\theta_W)^2}$
$TT\mathcal{HH}$	$\dfrac{(\sin\theta_T + \kappa \cos\theta_T)^2}{2(\cos\theta_W \sin\theta_W)^2}$
$TZ\mathcal{HH}$	$\dfrac{(\sin\theta_T + \kappa \cos\theta_T)(\kappa \sin\theta_T - \cos\theta_T)}{2(\cos\theta_W \sin\theta_W)^2}$
$ZZ\mathcal{SS}$	$\dfrac{(\tau \sin\theta_T)^2}{2(\cos\theta_W \sin\theta_W \tan\beta)^2}$
$TT\mathcal{SS}$	$\dfrac{(\tau \cos\theta_T)^2}{2(\cos\theta_W \sin\theta_W \tan\beta)^2}$
$TZ\mathcal{SS}$	$\dfrac{\tau^2 \sin\theta_T \cos\theta_T}{2(\cos\theta_W \sin\theta_W \tan\beta)^2}$

References

1. Weinberg, S. A Model of Leptons. *Phys. Rev. Lett.* **1967**, *19*, 1264–1266, doi:10.1103/PhysRevLett.19.1264. [CrossRef]
2. Aad, G.; Abajyan, T.; Abbott, B.; Abdallah, J.; Khalek, S.A.; Abdelalim, A.A.; Abdinov, O.; Aben, R.; Abi, B.; Abolins, M.; et al. Observation of a new particle in the search for the Standard Model Higgs boson with the ATLAS detector at the LHC. *Phys. Lett.* **2012**, *B716*, 1–29, doi:10.1016/j.physletb.2012.08.020. [CrossRef]
3. Chatrchyan, S.; Khachatryan, V.; Sirunyan, A.M.; Tumasyan, A.; Adam, W.; Aguilo, E.; Bergauer, T.; Dragicevic, M.; Erö, J.; Fabjan, C.; et al. Observation of a new boson at a mass of 125 GeV with the CMS experiment at the LHC. *Phys. Lett.* **2012**, *B716*, 30–61, doi:10.1016/j.physletb.2012.08.021. [CrossRef]
4. Tanabashi, M.; Hagiwara, K.; Hikasa, K.; Nakamura, K.; Sumino, Y.; Takahashi, F.; Tanaka, J.; Agashe, K.; Aielli, G.; Amsler, C.; et al. Review of Particle Physics. *Phys. Rev.* **2018**, *D98*, 030001, doi:10.1103/PhysRevD.98.030001. [CrossRef]
5. Standard Model Physics. 2019. Available online: https://twiki.cern.ch/twiki/bin/view/AtlasPublic/StandardModelPublicResults (accessed on 5 January 2020).
6. Summaries of CMS Cross Section Measurements. 2019. Available online: https://twiki.cern.ch/twiki/bin/view/CMSPublic/PhysicsResultsCombined (accessed on 5 January 2020).
7. Bezrukov, F.; Shaposhnikov, M. Standard Model Higgs boson mass from inflation: Two loop analysis. *J. High Energy Phys.* **2009**, *7*, 89, doi:10.1088/1126-6708/2009/07/089. [CrossRef]
8. Degrassi, G.; Di Vita, S.; Elias-Miro, J.; Espinosa, J.R.; Giudice, G.F.; Isidori, G.; Strumia, A. Higgs mass and vacuum stability in the Standard Model at NNLO. *J. High Energy Phys.* **2012**, *8*, 98, doi:10.1007/JHEP08(2012)098. [CrossRef]
9. Englert, F.; Brout, R. Broken Symmetry and the Mass of Gauge Vector Mesons. *Phys. Rev. Lett.* **1964**, *13*, 321–323, doi:10.1103/PhysRevLett.13.321. [CrossRef]
10. Higgs, P.W. Broken Symmetries and the Masses of Gauge Bosons. *Phys. Rev. Lett.* **1964**, *13*, 508–509, doi:10.1103/PhysRevLett.13.508. [CrossRef]
11. Schabinger, R.M.; Wells, J.D. A Minimal spontaneously broken hidden sector and its impact on Higgs boson physics at the large hadron collider. *Phys. Rev.* **2005**, *D72*, 093007, doi:10.1103/PhysRevD.72.093007. [CrossRef]
12. Pospelov, M.; Ritz, A.; Voloshin, M.B. Secluded WIMP Dark Matter. *Phys. Lett.* **2008**, *B662*, 53–61, doi:10.1016/j.physletb.2008.02.052. [CrossRef]

13. Basso, L.; Belyaev, A.; Moretti, S.; Shepherd-Themistocleous, C.H. Phenomenology of the minimal B-L extension of the Standard model: Z' and neutrinos. *Phys. Rev.* **2009**, *D80*, 055030, doi:10.1103/PhysRevD.80.055030. [CrossRef]
14. Appelquist, T.; Dobrescu, B.A.; Hopper, A.R. Nonexotic neutral gauge bosons. *Phys. Rev.* **2003**, *D68*, 035012, doi:10.1103/PhysRevD.68.035012. [CrossRef]
15. Duch, M.; Grzadkowski, B.; McGarrie, M. A stable Higgs portal with vector dark matter. *J. High Energy Phys.* **2015**, *9*, 162, doi:10.1007/JHEP09(2015)162. [CrossRef]
16. Péli, Z.; Trócsányi, Z. Stability of the vacuum as constraint on $U(1)$ extensions of the standard model. *arXiv* **2019**, arXiv:1902.02791.
17. Basso, L. Phenomenology of the Minimal B-L Extension of the Standard Model at the LHC. Ph.D. Thesis, University of Southampton, Southampton, UK, 2011.
18. Blennow, M.; Fernandez-Martinez, E. Parametrization of Seesaw Models and Light Sterile Neutrinos. *Phys. Lett.* **2011**, *B704*, 223–229, doi:10.1016/j.physletb.2011.09.028. [CrossRef]
19. Del Aguila, F.; Masip, M.; Perez-Victoria, M. Physical parameters and renormalization of U(1)-a x U(1)-b models. *Nucl. Phys.* **1995**, *B456*, 531–549, doi:10.1016/0550-3213(95)00511-6. [CrossRef]

© 2020 by the authors. Licensee MDPI, Basel, Switzerland. This article is an open access article distributed under the terms and conditions of the Creative Commons Attribution (CC BY) license (http://creativecommons.org/licenses/by/4.0/).

Article
Does Our Universe Prefer Exotic Smoothness?

Torsten Asselmeyer-Maluga [1,2,†], **Jerzy Król** [2,3,*,†] **and Tomasz Miller** [2]

1. German Aerospace Center (DLR), Rutherfordstr. 2, 12489 Berlin, Germany; Torsten.Asselmeyer-Maluga@dlr.de
2. Copernicus Center for Interdisciplinary Studies, Jagiellonian University, Szczepańska 1/5, 31-011 Cracow, Poland; tomasz.miller@uj.edu.pl
3. Cognitive Science and Mathematical Modelling Chair, University of Information Technology and Management, ul. Sucharskiego 2, 35-225 Rzeszów, Poland
* Correspondence: iriking@wp.pl
† These authors contributed equally to this work.

Received: 24 November 2019; Accepted: 2 January 2020; Published: 5 January 2020

Abstract: Various experimentally verified values of physical parameters indicate that the universe evolves close to the topological phase of exotic smoothness structures on $\mathbb{R}^4$ and K3 surface. The structures determine the α parameter of the Starobinski model, the number of e-folds, the spectral tilt, the scalar-to-tensor ratio and the GUT and electroweak energy scales, as topologically supported quantities. Neglecting exotic R^4 and K3 leaves these free parameters undetermined. We present general physical and mathematical reasons for such preference of exotic smoothness. It appears that the spacetime should be formed on open domains of smooth K3#$\overline{CP^2}$ at extra-large scales possibly exceeding our direct observational capacities. Such potent explanatory power of the formalism is not that surprising since there exist natural physical conditions, which we state explicitly, that allow for the unique determination of a spacetime within the exotic K3.

Keywords: exotic R^4 and cosmology; space topology changes; exotic K3; spacetime

1. Introduction

The micro-scale of the physical world and the large cosmological scales, when organised into a single cosmological model of the universe, should be finely interrelated. Even though we do not fully understand how these scales might intersect and interact with each other, our partial understanding allows for important insights. In particular, we expect that the complete picture of the domain of their common applicability would be a crucial ingredient of the successful theory of quantum gravity. The reason is simple: The universe at large scales where gravity dominates is described by the theory of general relativity (GR), whereas at the micro-scale the suitable theory is quantum mechanics (QM).

There are many reasons to introduce exotic smoothness. From the physics point of view one natural reason is quantum gravity. In the last years, we developed an approach, smooth quantum gravity, where the quantization procedure is given by a change of the smoothness structure [1]. The approach works only for four-dimensional spacetimes and has many connections to noncommutative geometry. Loosely speaking, the change of the smoothness structure is a quantization of the geometry in the sense of quantum gravity. A direct consequence of this approach is the determination of topology changes. To illustrate, let us consider a spacetime of topology $S^3 \times \mathbb{R}$. In the usual smoothness structure, this spacetime is foliated like $S^3 \times \{t\}$, i.e., the topology of the space S^3 remains constant. In contrast, a spacetime with topology $S^3 \times \mathbb{R}$ but exotic smoothness can also be foliated like $S^3 \times \{t\}$ but not smoothly. The smooth decomposition of an exotic $S^3 \times \mathbb{R}$ is a spacetime where the spatial component changes in a complicated process. Interestingly, the change seen as a process can be very different but the result of the change depends only on the topology of the spacetime. In the presented paper we construct a spacetime from first principles and show that there are two topology changes. Interestingly, this universal feature of spacetime can be understood by considering certain exotic $\mathbb{R}^4$.

The standard smoothness structure of $\mathbb{R}^4$ is the unique structure such that the product $\mathbb{R} \times \mathbb{R}^3$ is smooth. An exotic R^4 is a topological 4-manifold $\mathbb{R}^4$ which, if smooth, is nondiffeomorphic to the standard smooth $\mathbb{R}^4$. In any dimension other than 4, there exists a unique smoothness structure on $\mathbb{R}^n, n \neq 4$, the standard smooth $\mathbb{R}^n$. The existence of exotic R^4 was established in the 1980s and, together with the existence of at least two families of R^4s each containing uncountably infinitely many different nondiffeomorphic R^4s, are highly nontrivial mathematical facts (e.g., [2]). One such family of small exotic R^4s comprises those R^4s that are embeddable in the standard $\mathbb{R}^4$ as open subsets while the large exotic R^4s are not embeddable in $\mathbb{R}^4$ and hence in S^4.

The existence of such smooth exotic 4-manifolds may seem to be a purely mathematical curiosity; however, the application to physics also discussed in this paper shows it is not. On all (known) four-dimensional open manifolds there exist uncountably many different nondiffeomorphic smoothness structures. Compact 4-manifods can be endowed with countably many such structures. The main point advocated here and in our previous works is that one cannot understand the origins of certain values of important physical parameters (cosmology, particle physics) and one cannot understand the common domain of GR and quantum phenomena in the spacetime of dimension 4 without referring to exotic smooth 4-manifolds. Even though the current state of investigation does not support decisively and univocally the above categorical statements, the results collected strongly support them.

The exceptional (though quite direct) feature of exotic R^4s is that they are all Riemann smooth 4-manifolds which cannot be flat, i.e., their Riemann curvature tensors are not vanishing on any exotic R^4s. From the point of view of physics, a nonzero gravitational energy density is assigned to each exotic R^4, contrary to the case of the standard $\mathbb{R}^4$. Recently Gabor Etesi showed that certain smooth four-dimensional manifolds, namely the large exotic R^4s, are precisely the gravitational instantons [3]. Both these facts, being a Ricci-flat gravitational instanton and carrying nonzero gravitational energy, show that R^4s indeed place themselves in the overlapping domain of classical and quantum regimes of gravity. We will discuss the particular role played by the Ricci-flatness in the process of the generation of masses in spacetime. This is one of the first physical effects which has been considered in the context of exotic R^4 and it is known as the Brans conjecture. It states that exotic R^4s serve as sources of an external gravitational field in spacetime [4,5]. Moreover, R^4s determine noncommutative von Neumann algebras which is not the case for the standard $\mathbb{R}^4$ and this is yet another indication that R^4s are properly (though somewhat mysteriously) placed in the common domain of GR and QM (e.g., [1,6]). In recent publications [7–9] we have shown how the appearance of nonstandard smoothness on $\mathbb{R}^4$ and a K3 surface leads to explaining in purely topological terms the extremely tiny value of the cosmological constant and some other cosmological parameters.

This apparent multifaceted role of exotic smoothness on R^4 in physics, especially cosmology, motivates the attempt to understand the exotic smoothness as a consequence of certain, quite general, conditions imposed on physical spacetime of dimension 4. In what follows we explicitly state these conditions and discuss them from the physical and mathematical points of view. Both threads finely meet and intertwine in dimension 4 giving rise to a quite powerful explanatory framework. In particular it appears that considering space as homology 3-spheres (including S^3) is a general fact following from the causal and Lorentzian-metric structure (for a spacetime being a smooth 4-cobordism). Exotic 4-smoothness determines such cobordisms canonically which lies in the core of the presented approach. Finally, we overview and discuss the main results obtained within the framework.

2. Spacetime and Exotic Smoothness

In our previous work [9] we discussed a model with a compact spacetime, the K3 surface, where the cosmic evolution was given by an open submanifold. The important feature of the model is that a certain exotic R^4 is necessarily embedded into (a smooth version of) K3. Let us now reverse the argumentation and consider an evolution of the cosmos which starts with a 3-sphere and allows for spatial topology changes. As a consequence we will obtain the K3 surface with the two transitions as discussed in [9].

Topology describes the global properties of a manifold which are invariant with regard to the local shape or geometry. A local theory based on differential geometry like GR restricts very weakly the topology of spacetime. Because of this ambiguity as a rule we have to set a topology of the cosmos by hand, e.g., Einstein used the 3-sphere S^3 but $\mathbb{R}^3$ is another common choice.

Here we will discuss the topological implications of the assumed spacetime with an exotic smooth structure. We shall also need some further mild conditions to formulate a sufficiently useful cosmological model. The first condition is given by the measurement data of the cosmological background radiation of the COBE, WMAP and PLANCK experiments [10–13]. The analysis of the spectrum by Luminet et al. [14] gives a hint of a cosmos with a finite volume which is compatible with the Einstein cosmos S^3 or any other compact model, but not with $\mathbb{R}^3$. Thus our first condition on the topology of the cosmos is the following

1. The cosmos Σ is a compact 3-manifold without boundary.

Next we concentrate on spacetime. The choice of a spacetime is strongly restricted by two demands: Smoothability and causality (including the existence of a Lorentz metric). Usually the two conditions can be fulfilled if the spacetime M is diffeomorphic to $\Sigma \times \mathbb{R}$ with the (spatial) 3-manifold Σ, i.e., one makes the assumption that the topology of Σ is fixed. However, it is widely believed that the inclusion of quantum-gravitational effects enforces transitions of the (spatial) topology. We discussed in our previous works the possibility of an exotic smoothness structure which leads necessarily to topological transitions. To enable the topological transitions of Σ we have to model the spacetime as a cobordism M with $\partial M = \Sigma_0 \sqcup \Sigma$ describing the nontrivial evolution (i.e., $M \neq \Sigma \times \mathbb{R}$) from the initial state Σ_0 to the cosmos Σ at the epoch t. The cobordism M between a compact 3-manifold is also itself compact for a finite time interval. A compact manifold M possesses a Lorentz metric if (and only if) there exists a nonvanishing vector field, i.e., its Euler characteristic $\chi(M)$ is zero [15,16] or in case of the cobordism the relative Euler characteristic vanishes $\chi(M, \partial M) = 0$. Thus the second condition is:

2a. The relative Euler characteristic $\chi(M, \partial M)$ of the spacetime M is zero.

The topological censorship theorem [17] requires a simply connected spacetime. This is a necessary condition to avoid time-loops (which are contractible in a simply connected spacetime):

2b. The spacetime M is simply connected.

Conditions 2a and 2b imply the vanishing of the relative homology groups $H_k(M, \partial M) = 0$ for $k = 0, 2, 3$.

For let a 4-manifold M be 4-cobordism between two 3-manifolds Σ_1, Σ_2 such that $\partial M = \Sigma_1 \sqcup \Sigma_2$. To determine the homology of M, one has to use the following long exact sequence of homology groups

$$\ldots \to H_k(\partial M) \to H_k(M) \to H_k(M, \partial M) \to H_{k-1}(\partial M) \to H_{k-1}(M) \to \ldots$$

where the maps between the homology groups are induced by the inclusions $\partial M \to M$ and $M \to M/\partial M$. Now let us assume that M is simply connected, i.e., $H_1(M) = 0$ (Condition 2b above). We thus obtain the sequence

$$0 \to H_2(\partial M) \to H_2(M) \to H_2(M, \partial M) \to H_1(\partial M) \to 0 \tag{1}$$

where we used the Poincaré duality $H_3(M, \partial M) = H^1(M) = \text{Hom}(H_1(M), \mathbb{Z}) = 0$. For the other terms of the sequence we get $H_k(M, \partial M) = 0$ for $k = 0, 3$ (Betti numbers $b_0 = b_3 = 0$) and $H_k(M, \partial M) = \mathbb{Z}$ for $k = 1, 4$ (Betti numbers $b_1 = b_4 = 1$). In order to ensure the existence of a

Lorentz metric we need M to admit a nonvanishing time-like vector field which requires the relative Euler characteristics to vanish, $\chi(M, \partial M) = 0$. Since $\chi(M, \partial M) = b_0 - b_1 + b_2 - b_3 + b_4$, we obtain $\chi(M, \partial M) = b_2$. All in all, the demand that a Lorentz metric exists leads to $H_2(M, \partial M) = 0$. Therefore from Sequence (1) we obtain $H_1(\partial M) = 0 = H_2(\partial M)$ and hence the boundary ∂M must be a disjoint union of homology 3-spheres.

Thus we see that the physical conditions of the existence of a Lorentz metric (Condition 2a) and of causality (Condition 2b) are equivalent to the following condition for the topology of the cosmos:

3. The cosmos Σ is a homology 3-sphere.

Let us summarise the points above and draw conclusions for the entire spacetime M. Interestingly, the conditions stated above have a strong and direct connection to the smoothness structure of M. The spacetime M is assumed to be a 4-manifold with a metric fulfilling the Einstein equation and admitting a smoothness structure. The smoothness structure in dimension 4 is characterised by the embedding of a certain four-dimensional submanifold $A \subset M$ – the Akbulut cork. The Akbulut cork is a contractible 4-manifold with the boundary a homology 3-sphere [18]. Now we choose an exotic smoothness structure. This step is motivated by the generation of matter resulting from the exotic smoothness structure (see [19,20] for instance). The smoothness of the exotic M requires that the Akbulut cork of M possesses two homology 3-spheres as boundaries $\partial A = S_0 \sqcup S_1$ and that the initial sphere $S_0 = S^3$ is a simple 3-sphere contained in Σ_0 in agreement with the two physical conditions (2a and 2b) above. This is precisely the point where the exotic R^4 is generated: The neighbourhood of the Akbulut cork $N(A) \subset M$ as embedded in the 4-manifold M is an exotic R^4 if M admits an exotic smoothness structure (or M is exotic). Then, Conditions 1–3 lead us univocally to a simple cosmological model:

4. The spacetime M is a smooth 4-manifold with $\partial M = \Sigma_0 \sqcup \Sigma$, realising a cobordism between two homology 3-spheres.

Initial state: The cosmos begins as a compact 3-manifold Σ_0 without boundary (Condition 1) and possesses the topology of a homology 3-sphere (Condition 2).

Dynamics: The spacetime is a cobordism M with $\partial M = \Sigma_0 \sqcup \Sigma$ (Condition 3). This 4-manifold is simply connected (Condition 2b) and its pseudo-Riemannian metric (Condition 2a) is determined by the Einstein equation. The cosmos expands from Σ_0 to Σ with the scaling factor $a(t)$ determined by the Friedmann equation. It is interesting to note that cobordisms represent properly spacetime in the categorical approach by John Baez [21]. In Baez's representation the entire category of spacetime cobordisms (between 3-space manifolds) is considered leading to a natural connection with quantum mechanics (as in topological quantum field theory, TQFT). Even though in our approach the smoothness structures in dimension 4 determine nontrivial cobordisms and we do not discuss the quantum operator representation, still this would be an interesting nontrivial task to find connections with TQFT.

Topology transition: The homology of the cosmos is an invariant (both Σ_0 and Σ are homology 3-spheres, Conditions 2 and 3). The topology of the initial state Σ_0 may change to Σ by a homology-preserving transition (nontriviality of $M \neq \Sigma \times \mathbb{R}$).

In order to firmly establish the model we now have to choose tangible candidates for Σ_0 and Σ. One can exclude that Σ_0 is a point singularity because in this case we would have $\chi(M) = 1$ (i.e., the time-like vector field vanishes at this singular point). However, we have seen that the Akbulut cork of M is a cobordism between a 3-sphere S^3 and a homology 3-sphere S_1 and that $S^3 \subset \Sigma_0$. Thus, it seems natural to choose $\Sigma_0 = S^3$:

5. The initial state Σ_0 is the Einstein cosmos S^3.

This choice for the initial state is further supported by Ashtekar et al. [22] where the authors described a cosmological model with the big bounce effect (see also [23]). The model does not show a singularity, i.e., there is no big crunch but rather contraction is followed again by expansion.

The cork A with $\partial A = S^3 \sqcup S_1$ is a submanifold of M with $\partial M = S^3 \sqcup \Sigma$. Thus S_1 is in the interior of M and Σ is the boundary. Given Σ as the state at time t one can interpret S_1 as an intermediate state $\Sigma(t_1) = S_1$ with $t_1 < t$. However, according to Donaldson [24] not all homology 3-spheres are smoothly cobordant to S^3 (i.e., M with $\partial M = S^3 \sqcup \Sigma$ is not smooth for all Σ). There is no full classification of such homology 3-spheres but rather a long list of counterexamples. One example shows that there is no smooth cobordism M between S^3 and one or more Poincaré spheres. A large class of homology 3-spheres are Brieskorn spheres described as submanifolds of $\mathbb{C}^3$

$$\Sigma(a,b,c) = \left\{ (z_1, z_2, z_3) \in \mathbb{C}^4 \mid z_1^a + z_2^b + z_3^c = 0, |z_1|^2 + |z_2|^2 + |z_3|^2 = 1 \right\}$$

with a, b, c different prime numbers. The Brieskorn spheres are distinguished from other homology 3-spheres because they are irreducible and any homology 3-sphere is a sum of irreducible homology 3-spheres. Any irreducible 3-manifold Σ is characterized to be not splittable to the connected sum other than $\Sigma \# S^3$ (prime decomposition, see [25]), i.e., irreducible Σ can only be split trivially into $\Sigma \# S^3$ (diffeomorphic to Σ). Secondly, there is another splitting of irreducible 3-manifolds along 2-tori into simpler pieces, the so-called JSJ decomposition (Jaco–Shalen–Johannson decomposition, see [26]). The remaining pieces are called atoroidal irreducible 3-manifolds. Brieskorn spheres are the only nonhyperbolic irreducible homology 3-spheres. As we shall see shortly, these properties are crucial for applications in physics.

The solution of the geometrization conjecture implies that there are two important geometric classes of topological manifolds in dimension 3: Hyperbolic and nonhyperbolic 3-manifolds. The class of nonhyperbolic 3-manifolds is divided into seven subclasses among which there are the spherical and Euclidean geometries. Hyperbolic 3-manifolds are very special with respect to their properties. The main property important in this work is the rigidity of the volume for any diffeomorphism and conformal transformation (Mostow rigidity, see [27]), i.e., the volume is a topological invariant. Any scaling of a hyperbolic 3-manifold is an isometry or a hyperbolic 3-manifold cannot be scaled. This fact is extremely important for the evolution of the spatial component (as given by the cobordism M): If the intermediate state, say $\Sigma(t_1)$ at $t_0 < t_1 < t$, is a hyperbolic homology 3-sphere then the expansion of the spatial component has to stop (because of the Mostow rigidity). Therefore we have to assume that this intermediate state must be a nonhyperbolic 3-manifold. For simplicity reasons we choose an irreducible, nonhyperbolic 3-manifold (otherwise one has a sum of irreducible 3-manifolds as an intermediate earlier state which comprises of these irreducible 3-manifolds). For this reason the Brieskorn spheres are natural building blocks of all nonhyperbolic homology 3-spheres. The counterexample is the Poincaré sphere $\Sigma(2,3,5)$ which is the simplest one but cannot be used in any smooth cobordism with S^3. Moreover, the next one $\Sigma(2,3,7)$ provides another counterexample. The simplest Brieskorn sphere which is smoothly cobordant to S^3 is $\Sigma(2,5,7)$. Thus we look for an exotic M with the Akbulut cork A with $\partial A = S_0 \sqcup S_1$, $S_0 = S^3$ and $S_1 = \Sigma(2,5,7)$:

6. The intermediate state $\Sigma(t_1) = S_1$ at $t_0 < t_1 < t$ is the Brieskorn cosmos $\Sigma(2,5,7)$.

Finally we have to choose the 4-manifold M itself. There are two points of consideration which are important here. At first, in [19] we have shown that the transition of a standard 4-manifold to an exotic one results in non-Ricci-flatness. If we hypothesise that all matter terms in the Einstein–Hilbert action are only caused by exotic smoothness in the above way then the 4-manifold with its standard structure has to be Ricci-flat. However, there are only two compact 4-manifolds with a Ricci-flat metric, the 4-torus and the K3 surface

$$K = \left\{ (x,y,z,t) \in \mathbb{C}P^3 \mid x^4 + y^4 + z^4 + t^4 = 0 \right\}. \tag{2}$$

The 4-torus is a flat manifold that is not simply connected and so it contradicts Condition 2b, thus from the physical point of view the K3 surface is the preferred candidate of a spacetime. This is further supported by the second fact of consideration: The proposed 4-manifold A with $\partial A = S_0 \sqcup S_1$, $S_0 = S^3$ and $S_1 = \Sigma(2,5,7)$ is the Akbulut cork of a distinct 4-manifold K which is again the K3 surface.

The K3 surface is a compact 4-manifold with nonvanishing Euler characteristic and thus it admits no Lorentz metric. Therefore, the K3 surface itself cannot be the physical spacetime. However, we can imagine the cobordism M (with $\chi(M) = 0$ and equipped with a Lorentz metric) embedded in K. The submanifold $M \subset K$ is determined by K if one requires that both manifolds have the same Akbulut cork A with $\partial A = S_0 \sqcup S_1$, $S_0 = S^3$ and $S_1 = \Sigma(2,5,7)$. The choice $S_0 = S^3$ (Condition 4) is extended to the cork of the K3 surface if one replaces K by a version of the K3 surface $\mathcal{K} = K \setminus D^4$ with boundary $\partial(K \setminus D^4) = S^3$, i.e., we get $\partial \mathcal{K} = \partial(K \setminus D^4) = S^3 = S_0 = \Sigma_0$. Thus we arrive at the last condition of the model:

7. The K3 surface $\mathcal{K} = K \setminus D^4$ determines the 4-manifold M with $\partial M = S^3 \sqcup \Sigma$ by its common Akbulut cork. M is the physical spacetime.

Then the boundary component S^3 of M agrees with $\partial \mathcal{K}$ and M contains also the Akbulut cork A of $\mathcal{K}$, i.e., the 4-manifold representing the first transition $S_0 = S^3 \to S_1 = \Sigma(2,5,7)$ is the Akbulut cork of $\mathcal{K}$. Let us assume that the matter component in spacetime is caused by the exotic smoothness. However, the exotic smoothness is not determined by the topology of the Akbulut cork A but by the embedding of A into $\mathcal{K}$. Therefore we have to determine the neighbourhood $N(A) \subset \mathcal{K}$ of A in $\mathcal{K}$ to determine the smoothness structure. However, then the remaining part $\mathcal{K} \setminus N(A)$ is obtained purely by its topology. The boundary $\partial(\mathcal{K} \setminus N(A)) = S^3 \sqcup \Sigma$ contains the second component Σ (as a boundary of $N(A)$) which is also a homology 3-sphere (using the result of Freedman [18,28]). The topology of Σ is partly determined by the topology of $\mathcal{K}$. The reasons are the following.

Topological 4-manifolds are classified by the intersection form σ [18]. In case of our 4-manifold $\mathcal{K}$, one obtains

$$\sigma_\mathcal{K} = E_8 \oplus E_8 \oplus \begin{pmatrix} 0 & 1 \\ 1 & 0 \end{pmatrix} \oplus \begin{pmatrix} 0 & 1 \\ 1 & 0 \end{pmatrix} \oplus \begin{pmatrix} 0 & 1 \\ 1 & 0 \end{pmatrix}$$

$$= 2E_8 \oplus 3 \begin{pmatrix} 0 & 1 \\ 1 & 0 \end{pmatrix} = 2E_8 \oplus 3H$$

in the usual notation. The intersection form of the Akbulut cork A, as well as of $N(A)$, vanishes. By the splitting theorem in [29] one obtains

$$\sigma_{\mathcal{K} \setminus N(A)} = \sigma_\mathcal{K} = 2E_8 \oplus 3H$$

i.e., the same intersection form. Now $\mathcal{K} \setminus N(A)$ has the boundary

$$\partial(\mathcal{K} \setminus N(A)) = S^3 \sqcup \Sigma$$

and must be a smooth 4-manifold. Especially the block structure of the intersection form is reflected by the splitting of a 4-manifold. With these information we obtain the following general result

$$\Sigma = P \# P \# (K_1 \# K_2 \# K_3) \# S^3. \tag{3}$$

This 3-manifold Σ is also a homology 3-sphere consisting of three principal parts: The connected sum $P \# P$ of two Poincaré spheres, the connected sum of three irreducible homology 3-spheres $K_1 \# K_2 \# K_3$ and a 3-sphere. Of course one can omit the last 3-sphere but we keep it here as a reminder that the 3-sphere is always present in the connected sum $\#$ not changing the diffeomorphism class.

With Decomposition (3) at hand, we are able to complete our model using all six conditions above. It starts with a 3-sphere (Einstein cosmos), then the first transition to the Brieskorn sphere $\Sigma(2,5,7)$ takes place and finally it changes (second transition) to $\Sigma = P \# P \# (K_1 \# K_2 \# K_3) \# S^3$.

The two transitions are interpreted as inflationary phases [7,30] determining also the neutrino masses [8]. The three irreducible homology 3-spheres K_1, K_2, K_3 are identified with hyperbolic, homology 3-spheres inducing the matter part of the universe [19,20] with connections [6] to the models of Furey [31,32], Gresnigt [33], Bilson–Thompson [34,35]. The transition to the P#P part gives the cosmological constant [9]. Then, following the logic of the cosmological standard model, the remaining part S^3 (appearing as $S^2 \times [0,1]$ in the sum above) must be the dark matter component which will be discussed in a forthcoming paper. Finally we arrive at the picture:

- P#P causes the cosmological constant (= dark energy)
- K_1, K_2, K_3 is responsible for the matter part (= three generations?)
- S^3 or $S^2 \times [0,1]$ is associated with the dark matter (in the form of a gravitational soliton?)

3. Physical Parameters

Let us collect and discuss the results obtained on the base of our topological model of the evolving cosmos. Exotic smoothness in dimension 4 is the main player in the model. Therefore we have to motivate the appearance of exotic smoothness. The approach in the previous section is based extensively on the concept of cobordism for the spacetime. Thus, one has to consider the bounadry terms of the Einstein–Hilbert action. As starting point, let us discuss the Einstein–Hilbert action for a 4-manifold with boundary following our work [20]. In general, for a manifold M with boundary $\partial M = \Sigma$ one has the expression

$$S_{EH}(M) = \int_M R \sqrt{g}\, d^4x + \int_\Sigma H \sqrt{h}\, d^3x$$

where H is the mean curvature of the boundary with metric h. In the following we will discuss the boundary term, i.e., we reduce the problem to the discussion of the action

$$S_{EH}(\Sigma) = \int_\Sigma H \sqrt{h}\, d^3x \qquad (4)$$

along the boundary Σ (a 3-manifold). Following [20], Equation (4) over a 3-manifold Σ is equivalent to the Dirac action of a spinor over Σ. Main result of [20] is the following relation between the corresponding Dirac operators

$$D^M \Phi = D^\Sigma \psi - H\psi \qquad (5)$$

where D^Σ or D^M denote the Dirac operator on the 3-manifold Σ or 4-manifold, respectively. Now Φ must be a parallel spinor, i.e.,

$$D^M \Phi = 0 \qquad (6)$$

Finally we get

$$D^\Sigma \psi = H\psi \qquad (7)$$

leading to

$$\int_\Sigma H \sqrt{h}\, d^3x = \int_\Sigma \bar{\psi} D^\Sigma \psi \sqrt{h} d^3x \qquad (8)$$

In our previous work [1] we discussed a foliation of the 3-manifold which extends to the cobordism representing the topology change of the 3-manifold. There, we introduced the Godbillon–Vey invariant as topological invariant of the foliation. This foliation of codimension one is defined by a one-form ω (the leaves are the constant values) with integrability condition $\omega \wedge d\omega = 0$. Then the Godbillon–Vey invariant is defined by an integral over the 3-form $\eta \wedge d\eta$ with $d\omega = -\eta \wedge \omega$. Clearly, the foliation will also influence the spinor defined by Equation (8). For that purpose we reinterpret the invariant

$gv = \eta \wedge d\eta$ as the abelian Chern–Simons form for the abelian gauge field η. Then a covariant constant 1-form ω such that
$$D_\eta \omega = d\omega + \eta \wedge \omega = 0$$
defines a foliation, because the integrability condition $\omega \wedge d\omega = 0$ is automatically fulfilled. However, here we will use the coupling between the abelian gauge field η and the spinor ψ to the Dirac–Chern–Simons action functional on the 3-manifold
$$S_{DCS} = \int_\Sigma \left(\bar{\psi} D_\eta^\Sigma \psi \sqrt{h} d^3 x + \eta \wedge d\eta \right)$$
with the critical points at the solution
$$D_\eta^\Sigma \psi = 0 \quad d\eta = \tau(\psi, \psi)$$
where $\tau(\psi, \psi)$ is the unique quadratic form for the spinors locally given by $\bar{\psi}\gamma^\mu \psi$. Now we consider a spacetime $\Sigma \times I$, so that the solution is translationally invariant. Expressed differently, we choose a spacetime with foliation induced by the foliation of Σ extended by translation. An alternative description for this choice is by considering the gradient flow of these equations
$$\frac{d}{dt}\eta = d\eta - \tau(\psi, \psi)$$
$$\frac{d}{dt}\psi = D_\eta^\Sigma \psi$$

However, it is known that this system is equivalent to the Seiberg–Witten equation for $\Sigma \times I$ by using an appropriated choice of the so-called $Spin_C$ structure. Then this $Spin_C$ structure is directly related to the foliation. Therefore a nontrivial foliation together with the existence of Fermions induces a nontrivial solution of the gradient system which results in a nontrivial solution of the Seiberg–Witten equations. However, this nontrivial solution (i.e., $\psi \neq 0, \eta \neq 0$) is a necessary condition for the existence of an exotic smoothness structure.

With these arguments we obtained a strong relation between foliations, exotic smothness and our model for a spacetime (with spatial topology change). The origin of this foliation can be traced back to the Einstein cosmos. As discussed in [1], this initial state S^3 of the universe cannot be a smooth S^3 but rather a wild embedded 3-sphere (representing the quantum geometry of the quantum state). It is a direct consequence of exotic smoothness. As shown in the previous section, this initial state determines the stages of all further changes. In particular, it determines the growing of the 3-manifolds within the topology changing process. This process is related to hyperbolic geometry so that the scaling parameter a of the 3-manifold is part of the hyperbolic metric da^2/a^2 (relative to the scaling change $d\vartheta^2$ along the cobordism), i.e., we have the relation
$$\frac{da^2}{a^2} = d\vartheta^2$$
between the foliation of the wild embedded 3-sphere and the foliation of the cobordism (representing the topology change) leading to the formal solution
$$a = a_0 \exp(\vartheta).$$
This relation is at the root of the exponential behaviour for the physical parameters, e.g., the scaling parameter reads [9]
$$\vartheta = \frac{3}{2 \cdot CS(\Sigma)}.$$

Here the embedding of the exotic R^4 is important because it is directly related to the wild embedding of the 3-sphere representing the initial state. With the arguments above, one obtains an independent derivation of various results based on the exponential behaviour above. This shows that the model in the previous section is completely consistent with the previous work. For completeness, in what follows, we will present main results of this kind.

The curvature of an exotic R^4 depends on the embedding into a broader manifold. Still one can extract the invariant topological quantity of the curvature which corresponds to the embedding. The deep result of [9] is that one finds that the topological invariant quantity of the embedding $R^4 \to K3 \# \overline{CP^2}$ explains the tiny necessarily nonzero value of the cosmological constant (CC). Thus the value of CC is a topological invariant corresponding to the two topology changes as in the previous section, $S^3 \to \Sigma(2,5,7) \to P\#P$, and is given by the formula

$$\Omega_\Lambda = \frac{c^5}{24\pi^2 h G H_0^2} \exp\left(-\frac{3}{CS(\Sigma(2,5,7))} - \frac{3}{CS(P\#P)} - \frac{\chi(A)}{4}\right) \simeq 0.7029 \qquad (9)$$

where quantum corrections are included (represented by 1/4th part of the Euler characteristic of the Akbulut cork A [1] with $\partial A = \Sigma(2,5,7)$). $CS(\Sigma(2,5,7))$ and $CS(P\#P)$ are the Chern–Simons invariants of $\Sigma(2,5,7)$ and $P\#P$, respectively. Thus we have a topological scenario explaining the realistic value of CC avoiding the zero-point energies excessive contributions. The topological invariance does the job: Such a CC value is not an additive quantity since otherwise the topological invariance would be spoiled. We can understand this also by making use of smallness of exotic R^4s as follows.

The defining property of any small R^4 is its embedding into the standard $\mathbb{R}^4$. The invariant topological part of the cosmological constant in this case reads [9]

$$\text{(the CC of the embedding } R^4 \hookrightarrow \mathbb{R}^4) = \frac{1}{\sqrt[3]{\text{Vol}^2(Y_\infty)}} \exp\left(-\frac{3}{CS(Y_\infty)}\right) \qquad (10)$$

where Y_∞ is a 3-sphere widely embedded in R^4 with the volume $\text{Vol}(Y_\infty)$ and $CS(Y_\infty)$ is its Chern–Simons invariant. As noted in [9] the Chern–Simons invariant of such a sphere vanishes and so the value of CC vanishes as well by (10). This is a quite remarkable result by itself. Every small exotic R^4 is embeddable in $\mathbb{R}^4$ and the curvature of R^4 depends on the embedding. However, whatever values the Riemann curvature takes the invariant parts for the embedding are always zero. Thus the CC value vanishes for every small R^4 embedded in $\mathbb{R}^4$.

Consider a quantum field theory defined on the Minkowski spacetime M^4 and allow for the (quantum) fluctuations of curvature which lead to a Lorentzian spacetime manifold $\tilde{M}^4$. Even though we do not know the precise quantum description of gravitational fluctuations we still accept the point of view that in the semiclassical limit the zero-point energies of quantum fields give nonvanishing contributions to the vacuum energy density in spacetime. Is it possible that the curvature of $\tilde{M}^4$ be generated by smoothness structure on $\mathbb{R}^4$? Let us consider a certain exotic R^4. Since it is open we can always find a nonvanishing smooth vector field $X(x)$ on R^4 and define a curved Lorentzian manifold M_X^4 (e.g., [3]). This construction depends on $X(x)$ but since the embedding $R^4 \hookrightarrow \mathbb{R}^4$ varies the Riemann curvature of R^4, the curvature of the corresponding Lorentz manifold M_X^4 varies as well. Thus for such a class of Lorentzian manifolds which are of the form M_X^4 for some R^4 and a nonvanishing vector field X on it, the corresponding invariant value of CC vanishes. This can serve as a topological mechanism explaining the vanishing of CC on certain Lorentzian spacetime manifolds. However, the mechanism works under a supposition that the CC contributions on flat Minkowski spacetime generate the curvature which comes from the exotic R^4 as described above. This means that the vanishing of CC can be achieved via changing the smoothness from the standard $\mathbb{R}^4$ to the small exotic R^4 and subsequently considering embedding of the latter into the standard $\mathbb{R}^4$.

Thus we need a two-step extension of spacetime to understand the observed value of CC by topological means (any exotic R^4 is locally the standard $\mathbb{R}^4$): $\mathbb{R}^4 \to R^4$ and $R^4 \to K3$. In fact this kind of a topological approach is quite universal and a couple of other cosmological parameters can be similarly derived as topological invariants. The following examples show the scope of the approach [7,8].

1. The α parameter in the Starobinsky model (in the units of the Planck mass squared)

$$\alpha \cdot M_P^{-2} = \frac{1}{\left(1 + \vartheta + \frac{\vartheta^2}{2} + \frac{\vartheta^3}{6}\right)} \approx 10^{-5} \text{ where } \vartheta = \frac{3}{2 \cdot CS(\Sigma(2,5,7))} = \frac{140}{3}.$$

2. The number of e-folds during the inflation

$$N = \frac{3}{2 \cdot CS(\Sigma(2,5,7))} + \ln(8\pi) \approx 51.$$

3. The scalar/tensor ratio $r = \frac{12}{(\vartheta + \ln(8\pi^2))^2} \approx 0.0046$.

4. The spectral tilt $n_s = 1 - \frac{2}{\vartheta + \ln(8\pi^2)} \approx 0.961$.

5. The GUT energy scale (the energy of the first topology change $S^3 \to \Sigma(2,5,7)$)

$$\Delta E_1 = \frac{E_{\text{Planck}}}{1 + \vartheta + \frac{\vartheta^2}{2} + \frac{\vartheta^3}{6}} \approx 10^{15} \text{ GeV}.$$

6. The electroweak energy scale (the energy assigned to the second topological transition $\Sigma(2,5,7) \to P\#P$)

$$E_2 = \frac{E_{\text{Planck}} \cdot \exp\left(-\frac{1}{2 \cdot CS(P\#P)}\right)}{1 + \vartheta + \frac{\vartheta^2}{2} + \frac{\vartheta^3}{6}} \approx 63 \text{ GeV}.$$

7. The topological bound on the sum of the three neutrino masses < 0.018 eV.

Together with the value of CC the above list strongly suggests that the topology underlying exotic smooth 4-manifolds, like R^4 and K3, might indeed shed some light on the important domains of physics where certain crucial physical parameters remain free or theoretically undetermined. This property of being topological invariant with respect to physical quantities indicates a fundamental character of the approach.

Is there any fundamental symmetry leading to topologically supported physical parameters? One indication follows from the constructions presented in this paper. Firstly, as presented in Section 2 the 4-cobordism between S^3 and $\Sigma(2,5,7)$ is a driving force for the smooth evolution of the cosmos and it yields the cosmological inflation with the realistic e-fold number and the value of the α parameter. The smoothness of such an evolution is restored as soon as one refers to the modified (exotic) smoothness on $\mathbb{R}^4$. The entire modification is caused by the Akbulut cork with the boundary $S^3 \sqcup \Sigma(2,5,7)$ and its embedding into R^4. This suggests that diffeomorphisms invariance in dimension 4 is somehow replaced by broader cobordisms invariance. Secondly, in order to understand the role of cobordisms between 4-manifolds let us start with recalling the following h-cobordism theorem in dimensions greater or equal to 6.

Let W be a simply connected compact manifold with a boundary ∂M that has two components, M_1 and M_2 such that the inclusions $i_{1,2}: M_{1,2} \hookrightarrow M$ are homotopy equivalences. Then W is diffeomorphic to the product $M_1 \times [0,1] = M_2 \times [0,1]$, where dimensions of $M_{1,2} \geq 5$. This means that if M_1 and M_2 are two simply connected manifolds of dimension ≥ 5 and there exists an h-cobordism W between them, then W is a product $M_1 \times [0,1]$ and M_1 is diffeomorphic to M_2.

In dimension 5, however, the following holds.

There exist simply connected compact cobordisms W of dimension 5 with the inclusions of their boundary components $M_{1,2} \stackrel{i_{1,2}}{\hookrightarrow} W$ being homotopy equivalences such that W is not diffeomorphic to the product $M_1 \times [0,1]$ (or $M_2 \times [0,1]$) hence M_1 is not diffeomorphic to M_2 being h-cobordant to it.

Thus there exists a five-dimensional smooth cobordism between nondiffeomorphic 4-manifolds which is topologically trivial. This phenomenon indicates that something unusual is happening in dimension 4 and in fact there follows the existence of small exotic R^4s. In the particular case $M_1 = 3CP^2 \# 20\overline{CP^2}$ and $M_2 = K3 \# \overline{CP^2}$, which are homeomorphic (and certainly homotopy equivalent) but not diffeomorphic, the 5-cobordism $W^5 = M_1 \times [0,1]$ is topologically trivial but smoothly nontrivial. Moreover, there exists the same Akbulut cork as considered previously: $\tilde{A} \subset M_1$ and $A \subset M_2$ ($\tilde{A}$ and A differ by a certain involution of the boundary ∂A) such that the neighbourhoods $N(\tilde{A})$ of $\tilde{A}$ in M_1 and $N(A)$ of A in M_2 are both (different) exotic R^4s. $N(A)$ in $K3 \# \overline{CP^2}$ is precisely the exotic R^4 leading to the realistic value of the cosmological constant and which has been referred to in this paper. Consequently, the Akbulut cork A realises the 4-cobordism between S^3 and $\Sigma(2,5,7)$ as described in Section 2.

Thus, four-dimensional nondiffeomorphic smooth manifolds R^4s in M_1 and M_2, and the possibility to attain one of them from the other via a nontrivial 5-cobordism, appears as the fundamental 'symmetry' of a physical theory extending GR. However, here we do not investigate this interesting point any further. It will be addressed in the future work.

Author Contributions: Conceptualization, T.A.-M. and J.K.; methodology, T.A.-M. and J.K.; investigation, T.A.-M., J.K. and T.M.; resources, T.A.-M., J.K. and T.M.; writing—original draft preparation, T.A.-M. and J.K.; writing—review and editing, T.M.; supervision, T.A.-M. and J.K. All authors have read and agree to the published version of the manuscript.

Acknowledgments: The authors appreciate comments and requirements made by referees. The corrections improved readability of the paper.

Conflicts of Interest: The authors declare no conflict of interest.

References

1. Asselmeyer-Maluga, T. Smooth quantum gravity: Exotic smoothness and quantum gravity. In *At the Frontier of Spacetime*; Asselmeyer-Maluga, T., Ed.; Fundamental Theories of Physics vol 183; Springer: Basel, Switzerland, 2016; pp. 247–308.
2. Gompf, R.; Stipsicz, A. *4-Manifolds and Kirby Calculus*; AMS: Providence, RI, USA, 1999.
3. Etesi, G. Strong cosmic censorship and topology change in four dimensional gravity. *arXiv* **2019**, arXiv:1905.03952.
4. Asselmeyer, T. Generation of source terms in general relativity by differential structures. *Class. Quant. Grav.* **1997**, *14*, 749–758. [CrossRef]
5. Sładkowski, J. Gravity on exotic R^4's with few symmetries. *Int. J. Mod. Phys. D* **2001**, *10*, 311–313. [CrossRef]
6. Asselmeyer-Maluga, T. Braids, 3-manifolds, elementary particles: Number theory and symmetry in particle physics. *Symmetry* **2019**, *11*, 1298. [CrossRef]
7. Asselmeyer-Maluga, T.; Krol, J. A topological model for inflation. *arXiv* **2018**, arXiv:1812.08158.
8. Asselmeyer-Maluga, T.; Krol, J. A topological approach to neutrino masses by using exotic smoothness. *Mod. Phys. Lett. A* **2019**, *34*, 1950097. [CrossRef]
9. Asselmeyer-Maluga, T.; Król, J. How to obtain a cosmological constant from small exotic R^4. *Phys. Dark Universe* **2018**, *19*, 66–77. [CrossRef]
10. Ade, P.A.; Aghanim, N.; Armitage-Caplan, C.; Arnaud, M.; Ashdown, M.; Atrio-Barandela, F.; Aumont, J.; Baccigalupi, C.; Banday, A.J.; Barreiro, R.B.; et al. Planck 2013 results. XVI. Cosmological parameters. *arXiv* **2013**, arXiv:1303.5076.

11. Ade, P.A.; Aghanim, N.; Arnaud, M.; Ashdown, M.; Aumont, J.; Baccigalupi, C.; Banday, A.J.; Barreiro, R.B.; Bartlett, J.G.; Bartolo, N.; et al. Planck 2015 results. XIII cosmological parameters. *Astron. Astrophys.* **2016**, *594*, A13.
12. Komatsu, E.E.; Smith, K.M.; Dunkley, J.; Bennett, C.L.; Gold, B.; Hinshaw, G.; Jarosik, N.; Larson, D.; Nolta, M.R.; Page, L.; et al. Seven-year Wilkinson Microwave Anisotropy Probe (WMAP) observations: Cosmological interpretation. *Astrophys. J. Suppl.* **2011**, *192*, 18. [CrossRef]
13. Roukema, B.F.; Bulinski, Z.; Szaniewska, A.; Gaudin, N.E. The optimal phase of the generalised Poincaré dodecahedral space hypothesis implied by the spatial cross-correlation function of the WMAP sky maps. *Astron. Astrophys.* **2008**, *486*, 55–74. [CrossRef]
14. Luminet, J.-P.; Weeks, J.R.; Riazuelo, A.; Lehoucq, R.; Uzan, J.-P. Dodecahedral space topology as an explanation for weak wide-angle temperature correlations in the cosmic microwave background. *Nature* **2003**, *425*, 593–595. [CrossRef]
15. Hawking, S.W.; Ellis, G.F.R. *The Large Scale Structure of Space-Time*; Cambridge University Press: Cambridge, UK, 1994.
16. Steenrod, N. *Topology of Fibre Bundles*; Princeton University Press: Princeton, NJ, USA, 1999.
17. Friedman, J.L.; Schleich, K.; Witt, D.M. Topological censorship. *Phys. Rev. Lett.* **1993**, *71*, 1486–1489. [CrossRef]
18. Freedman, M.H. The topology of four-dimensional manifolds. *J. Diff. Geom.* **1982**, *17*, 357–454. [CrossRef]
19. Asselmeyer-Maluga, T.; Rosé, H. On the geometrization of matter by exotic smoothness. *Gen. Rel. Grav.* **2012**, *44*, 2825–2856. [CrossRef]
20. Asselmeyer-Maluga, T.; Brans, C.H. How to include fermions into general relativity by exotic smoothness. *Gen. Relativ. Grav.* **2015**, *47*, 30. [CrossRef]
21. Baez, J.C. Higher-dimensional algebra and Planck scale physics. In *Physics Meets Philosophy at the Planck Scale*; Callender, C., Huggett, N., Eds.; Cambridge University Press: Cambridge, UK, 2001; pp. 177–195.
22. Ashtekar, A.; Pawlowski, T.; Singh, P.P. Quantum nature of the big bang. *Phys. Rev. Lett.* **2006**, *96*, 141301, [CrossRef] [PubMed]
23. Ashtekar, A.; Singh, P. Loop quantum cosmology: A status report. *Class. Quant. Grav.* **2011**, *28*, 213001, [CrossRef]
24. Donaldson, S.K. An application of gauge theory to the topology of 4-manifolds. *J. Diff. Geom.* **1983**, *18*, 269–316. [CrossRef]
25. Milnor, J. A unique decomposition theorem for 3-manifolds. *Amer. J. Math.* **1962**, *84*, 1–7. [CrossRef]
26. Jaco, W.; Shalen, P. Seifert fibered spaces in 3-manifolds. *Geom. Topol.* **1979**, 91–99. [CrossRef]
27. Mostow, G.D. Quasi-conformal mappings in n-space and the rigidity of hyperbolic space forms. *Publ. Math. IHES* **1968**, *34*, 53–104. [CrossRef]
28. Freedman, M.H. A fake $S^3 \times R$. *Ann. of Math.* **1979**, *110*, 177–201. [CrossRef]
29. Freedman, M.H.; Taylor, L. λ splitting 4-manifolds. *Topology* **1977**, *16*, 181–184. [CrossRef]
30. Asselmeyer-Maluga, T.; Król, J.J. Inflation and topological phase transition driven by exotic smoothness. *Adv. HEP* **2014**, *2014*, 867460. [CrossRef]
31. Furey, C. Charge quantization from a number operator. *Phys. Lett. B* **2015**, *742*, 195–199, arXiv:1603.04078.
32. Furey, C. Standard Model Physics from an Algebra? Ph.D. Thesis, University of Waterloo, Waterloo, ON, USA, 2015.
33. Gresnigt, N.G. Braids, normed division algebras, and standard model symmetries. *Phys. Lett. B* **2018**, *783*, 212–221. [CrossRef]
34. Bilson-Thompson, S.O. A topological model of composite preons. *arXiv* **2005**, arXiv:hep-ph/0503213v2.
35. Bilson-Thompson, S.O.; Markopoulou, F.; Smolin, L. Quantum gravity and the standard model. *Class. Quant. Grav.* **2007**, *24*, 3975–3994. [CrossRef]

© 2020 by the authors. Licensee MDPI, Basel, Switzerland. This article is an open access article distributed under the terms and conditions of the Creative Commons Attribution (CC BY) license (http://creativecommons.org/licenses/by/4.0/).

Article

Axion Stars

Hong Zhang

Physik-Department T31, Technische Universität München, 85748 Garching, Germany; hong.zhang@tum.de

Received: 4 November 2019; Accepted: 13 December 2019; Published: 20 December 2019

Abstract: The dark matter particle can be a QCD *axion* or axion-like particle. A locally over-densed distribution of axions can condense into a bound Bose–Einstein condensate called an *axion star*, which can be bound by self-gravity or bound by self-interactions. It is possible that a significant fraction of the dark matter axion is in the form of axion stars. This would make some efforts searching for the axion as the dark matter particle more challenging, but at the same time it would also open up new possibilities. Some of the properties of axion stars, including their emission rates and their interactions with other astrophysical objects, are not yet completely understood.

Keywords: axion stars; Bose stars; oscillons

1. Introduction

The *QCD axion* is one of the best motivated dark matter particle candidates, since it provides a solution to the QCD strong CP problem. (For a recent review, see [1].) The QCD axion is a boson with spin-0. It has a tiny mass and extremely weak couplings with the Standard Model particles, as well as extremely weak self-interactions. However, axion dark matter is not simple because axions are identical bosons. Its tiny mass indicates that if a large proportion of dark matter is axions, the occupation numbers can be very large. Therefore, the axions can form a *Bose–Einstein condensate* (BEC). The collective behavior of BEC can be very different from an ideal gas of bosons. The axion BEC can be bound gravitationally, which are called *axion stars*, or bound by self-interactions, which are called *axitons*. (For a recent review, see [2].) If a large fraction of the axion dark matter is in such bound configurations, the theoretical predictions of the behavior of dark matter could be dramatically different, which would affect the experimental searches.

The QCD axion has been strongly constrained [1]. The allowed range of axion mass has been reduced to between 10^{-6} and 10^{-2} eV. Axion dark matter can also more generally refer to other light spin-0 boson with a periodic potential for self-interaction. There are motivations from string theory and astrophysics for a dark matter particle that is a very light boson with mass as light as 10^{-22} eV [3–5]. In this proceeding, we focus mainly on the QCD axion, but many of our results are presented in a form that can be applied to other axion-like particles straightforwardly.

2. Axion Field Theories at Different Energy Scales

The fundamental quantum field theory for the QCD axion is an extension of the Standard Model with Peccei–Quinn (PQ) $U(1)$ symmetry. PQ symmetry is spontaneously broken by the ground state of a complex Lorentz–scalar field [6–8]. After symmetry breaking, the minima of the potential are a circle of radius f_a, which is called the *axion decay constant*. At momentum scales of order f_a, the axion field is the Goldstone mode corresponding to excitation of the scalar field along that circle.

The axion can be described by a field theory with a real Lorentz–scalar field $\phi(x)$ at momentum scales much smaller than f_a. Its potential must have the shift symmetry with $\phi(x) = \phi(x) + 2\pi f_a$.

When the energy scale is below the week scale, which is about 100 GeV, the interactions between axions and the Standard Model particles are:

$$\frac{\alpha_s}{8\pi f_a} \phi\, G^a_{\mu\nu} \tilde{G}^{a\mu\nu} + \frac{c_{\gamma 0}\alpha}{8\pi f_a} \phi\, F_{\mu\nu} \tilde{F}^{\mu\nu} + \frac{1}{2 f_a} J^\mu \partial_\mu \phi, \qquad (1)$$

where $G^a_{\mu\nu}$ and $F_{\mu\nu}$ are the QCD and QED field strengths, $\tilde{G}^a_{\mu\nu} = \tfrac{1}{2}\epsilon_{\mu\nu\lambda\sigma} G^{a\lambda\sigma}$ and $\tilde{F}_{\mu\nu}$ are the corresponding dual field strengths, and the current J^μ is a linear combination of axial-vector fermion currents. The specific value of $c_{\gamma 0}$ and the form of J^μ depend on the specific axion model. The QCD field-strength term in Equation (1) is proportional to the topological charge density $\alpha_s G^a_{\mu\nu} \tilde{G}^{a\mu\nu}/8\pi$. The shift symmetry of ϕ is guaranteed by the quantization of the QCD topological charge in Euclidean field theory.

When the momentum scale is further below the scale of QCD confinement, which is about 1 GeV, the gluon degree of freedom is replaced by the degree of freedoms of hadrons. Then, the axion self-interactions from the coupling to the gluon field in Equation (1) can be described by a real potential $V(\phi)$:

$$\mathcal{L} = \tfrac{1}{2}\partial_\mu \phi\, \partial^\mu \phi - V(\phi). \qquad (2)$$

The invariance of the Lagrangian under the shift symmetry $\phi(x) \to \phi(x) + 2\pi f_a$ requires the potential $V(\phi)$ to be a periodic function of ϕ: $V(\phi) = V(\phi + 2\pi f_a)$. The Lagrangian is also invariant under the Z_2 symmetry $\phi(x) \to -\phi(x)$, which requires $V(\phi)$ to be an even function of ϕ.

The potential $V(\phi)$ for the axion field is determined by nonperturbative effects of QCD. The specific form can be systematically derived order-by-order from the chiral effective field theory for light pseudoscalar mesons of QCD and the axion [9]. The leading order derived from the chiral effective field theory for the axion and pions gives the *chiral potential* [10]:

$$V(\phi) = (m_\pi f_\pi)^2 \left(1 - \frac{\sqrt{1 + z^2 + 2z \cos(\phi/f_a)}}{1 + z}\right), \qquad (3)$$

where $z = m_u/m_d$ is the ratio of the up quark and down quark masses. The coefficient can be calculated by the pion mass $m_\pi = 135.0$ MeV and the pion decay constant $f_\pi = 92.2$ MeV, which are related to m_a and f_a by [11]

$$m_\pi f_\pi = \frac{1+z}{\sqrt{z}} m_a f_a. \qquad (4)$$

A next-to-leading order analysis in the chiral effective field theory gives the numerical value $z = 0.48(3)$ [9]. With the upper and lower bounds on f_a from cosmology and astrophysics, the allowed mass range for the QCD axion is between 6×10^{-6} and 2×10^{-3} eV [1]. In this proceeding, every time we provide a numerical value which depends on m_a, we give the value in the form:

$$m_a = 10^{-4 \pm 1} \text{eV}. \qquad (5)$$

It should be understood, as the value is between 10^{-5} and 10^{-3} depending on the choice of the axion mass. In addition to the more precise chiral potential, a popular model for the axion potential that has been widely used in phenomenological studies is called the *instanton potential*:

$$V(\phi) = (m_a f_a)^2 [1 - \cos(\phi/f_a)]. \qquad (6)$$

It can be derived with a dilute instanton gas approximation [12], which cannot be improved systematically. The field theory given by the Lagrangian in Equation (2) with the instanton potential in Equation (6) is often called the *sine-Gordon model*.

The axion field can be more simply expressed as a complex scalar field in a nonrelativistic effective field theory (NREFT) when the energy scale is much smaller than the axion mass m_a. A naive way of deriving the NREFT is to replace the real field by the complex field ψ with:

$$\phi(r,t) \approx \frac{1}{\sqrt{2m_a}} \left(\psi(r,t) e^{-im_a t} + \psi^*(r,t) e^{+im_a t} \right). \tag{7}$$

Then, by dropping out the terms with a rapidly oscillating phase in the form of $\exp(inm_a t)$ with nonzero integer n, we can get the NREFT Lagrangian:

$$\mathcal{L}_{\text{eff}} = \tfrac{1}{2} i \left(\psi^* \dot\psi - \dot\psi^* \psi \right) - \mathcal{H}_{\text{eff}}. \tag{8}$$

This effective Hamiltonian density depends on the field ψ and its gradients. It can be separated into three parts: $\mathcal{H}_{\text{eff}} = \mathcal{T}_{\text{eff}} + V_{\text{eff}} + W_{\text{eff}}$, where $\mathcal{T}_{\text{eff}}$ is the kinetic energy density, V_{eff} is a function of $\psi^*\psi$ only, and W_{eff} consists of all other interaction terms that also depend on gradients of ψ. An n-body term in $\mathcal{H}_{\text{eff}}$ has n factors of ψ and n factors of ψ^*, with arbitrary numbers of gradients. The kinetic energy density $\mathcal{T}_{\text{eff}}$ includes all the one-body terms:

$$\mathcal{T}_{\text{eff}} = \frac{1}{2m_a} \nabla \psi^* \cdot \nabla \psi - \frac{1}{8m_a^3} \nabla^2 \psi^* \nabla^2 \psi + \dots. \tag{9}$$

These terms reproduce the energy–momentum relation $E = \sqrt{m^2 + p^2} - m$ in the nonrelativistic limit. The effective potential V_{eff} can be expanded in powers of $\psi^*\psi$ beginning at order $(\psi^*\psi)^2$:

$$V_{\text{eff}}(\psi^*\psi) = m_a^2 f_a^2 \sum_{n=2}^{\infty} \frac{v_n}{(n!)^2} \left(\frac{\psi^*\psi}{2m_a f_a^2} \right)^n. \tag{10}$$

This NREFT Lagrangian can also be derived strictly by a nonlocal canonical transformation from the relativistic real scalar Lagrangian in Equation (2) [13]. The coefficients v_n are complex numbers in general, which can be calculated by matching the scattering amplitudes at low energy [14–16]. The contributions from Feynman diagrams without an internal propogator can be summed into a compact form, which is:

$$V_{\text{eff}}^{(0)}(\psi^*\psi) = (m_\pi f_\pi)^2 \left(1 - \frac{z}{4(1+z)^2} \hat{n} - \frac{1}{1+z} \int_0^1 dt \sqrt{1 + z^2 + 2z \cos(\hat{n}^{1/2} \sin(\pi t))} \right), \tag{11}$$

for chiral potential, and:

$$V_{\text{eff}}^{(0)}(\psi^*\psi) = (m_a f_a)^2 \left[1 - \tfrac{1}{4}\hat{n} - J_0(\hat{n}^{1/2}) \right]. \tag{12}$$

for instanton potential, where $\hat{n} = 2\psi^*\psi/(m_a f_a^2)$ is the dimensionless number density. A systematic scheme of including the off-shell internal propagators is suggested in [14].

3. Axion Stars

3.1. Dilute Axion Stars

An *axion star* is a boson star made of axions. The boson star with bosons in BEC was first considered by Tkachev [17]. The classical solutions for a boson star can be obtained by solving the Einstein–Klein–Gordon equations for a real scalar field $\phi(r,t)$ with axion potential $V(\phi)$. The solutions are approximately localized and periodic.

Stable solutions exist with the energy density at the center much lower than the QCD scale. We call these solutions *dilute axion stars*. The solutions of dilute axion stars can also be obtained using the axion NREFT with Newtonian gravity. The latter is much simpler and without loss of much accuracy:

$$i\dot\psi = -\frac{1}{2m_a}\nabla^2\psi + \left[V'_{\text{eff}}(\psi^*\psi) + m_a\Phi\right]\psi, \tag{13a}$$

$$\nabla^2\Phi = 4\pi G m_a \psi^*\psi. \tag{13b}$$

These equations are also called *Gross–Pitaevskii–Poisson (GPP) equations*. The number density $\psi^*\psi$ is much smaller compared to $m_a f_a^2$ for dilute axion stars. Thus, we can expand the effective potential and keep only the leading term:

$$V_{\text{eff}}(\psi^*\psi) \approx \frac{v_2}{16 f_a^2}(\psi^*\psi)^2. \tag{14}$$

Chavanis used the GPP equations with this leading term of the potential to derive simple approximations of the basic properties of boson stars with negative v_2 [18]. His results show that there is a maximum mass for the dilute axion stars $G^{-1/2} f_a/m_a$.

Variational methods has been used to get simple approximations to the dilute axion stars [18–20]. One can also match asymptotic expansions to get more accurate solutions [21]. the numerical method gives the most accurate solutions. The results below were obtained by numerically solving the GPP equations in Equation (13). The potential V_{eff} is either the naive effective instanton potential in Equation (12) or the naive effective chiral potential in Equation (11) with $z = 0.48$.

In Figure 1, we show the dependence of the radius R_{99} of the dilute axion star on the mass M. The critical point (indicated by the solid dot) separates the stable branch from the unstable branch. When the number density at the center n_0 increases, the solution moves from the left to the critical point along the stable branch. The dilute axion star at the critical point has the largest mass. Then, as n_0 keeps increasing, the solution moves from the critical point to the left along the unstable branch. The self-interaction of axions can be ignored for the stable solutions with a mass which is much smaller than the maximum mass. The self-interaction only plays an important role close to the critical point.

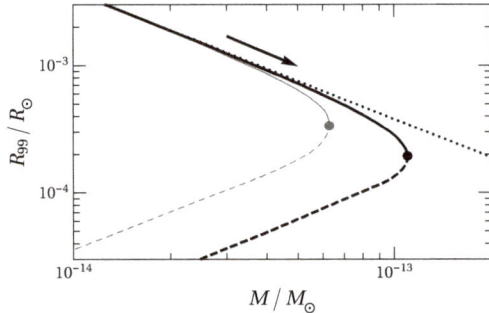

Figure 1. Radius R_{99} versus mass M for dilute axion stars. The axion mass is $m_a = 10^{-4}$ eV. The curves are calculated with chiral potential with $z = 0.48$ (black curves) or the instanton potential (gray curves). The dots are critical points at which the dilute axion stars have the maximum masses. The dots separate the unstable branch (dashed curve) from the stable branch (solid curve). For comparison, the boson stars with no self-interaction are shown with a dotted line. The arrow indicates the increase of axion star mass from the condensation of additional axions in the surroundings.

The properties of the critical point in Figure 1 are important in phenomenology. The number of axions in the dilute axion star with the chiral potential with $z = 0.48$ is $N_* = 1.2 \times 10^{57\mp3}$ for axion mass $m_a = 10^{-4\pm1}$ eV. (This number is smaller by the factor 0.59 for instanton potential because of the different value of v_2.) The corresponding critical dilute axion star is $N_* m_a = 1.1 \times 10^{-13\mp4} M_\odot$, with $M_\odot$ as the mass of the Sun. The critical radius $R_{99*} = 1.9 \times 10^{-4} R_\odot$, with $R_\odot$ as the radius of the Sun. More properties of the dilute axion stars can be found in [2].

3.2. Dense Axion Stars

In [22], we pointed out that there could be other stable branches of axion star solutions with larger center density. Another branch can be found by following the unstable solution from the critical dilute axion stars in Figure 1. With a larger center density, at some point, we need to consider all terms in the expansion of the potential $V_{\text{eff}}(\psi^*\psi)$. In [22], we solved the field equation in Equation (13) with the naive effective instanton potential in Equation (12). In [2], the results using the naive effective chiral potential in Equation (11) are obtained, which are also shown in Figure 2. A second critical point was found with the radius smaller by 7 orders of magnitude. The localized solutions near and beyond the second critical point were called the *dense axion stars* in [22], because the mass density $m_a \psi^* \psi$ at the center of the axion star becomes comparable to the QCD scale $(m_a f_a)^2$.

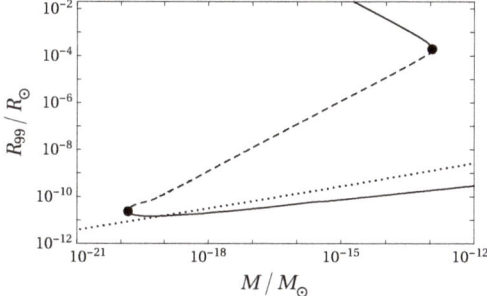

Figure 2. Radius R_{99} versus mass M for axion stars. The solutions are obtained with $m_a = 10^{-4}$ eV and the naive chiral potential with $z = 0.48$. The stable branches (solid curves) and unstable branch (dashed curves) are separated by critical points (labeled with black dots). The upper-left critical point is the same as the black dot in Figure 1. A second critical point is found by following the unstable solution to larger center density. Also shown is the Thomas–Fermi approximation (dotted curve).

For the dense axion stars near the second critical point in Figure 2, the contribution of gravity is almost negligible. Thus, the dense axion stars near the second critical point are actually oscillons. The oscillons are approximately localized solutions of a real scalar field, which are bound only by self-interactions. For the chiral potential with $m_a = 10^{-4\pm1}$ eV and $z = 0.48$, the critical number of axions is $N_* = 2 \times 10^{50\mp4}$. The critical mass $N_* m_a$ is $3 \times 10^{10\mp3}$ kg, and the critical radius R_{99*} is $2 \times 10^{-2\mp1}$ m. More properties of the critical dense axion stars can be found in [2].

As shown in Figure 2, beyond the lower critical point, the mass M of the dense axion star increases as a function of the radius R_{99}. With larger central density, the dense axion star curve approaches the Thomas–Fermi approximation [23], which is the straight dotted line in Figure 2. In the Thomas–Fermi approximation, the kinetic pressure is ignored except on the surface of the stars; in the bulk, it is the repulsive force from axion self-interaction which balances the attractive force from gravity. In [22], the Thomas–Fermi approximation was mistakenly used to extrapolate the curve of R_{99} to very large values of M. As pointed out in [24], the curve for R_{99} versus M actually crosses the line of the Thomas–Fermi approximation at a small angle. Therefore, the Thomas–Fermi approximation is not a proper estimate for dense axion stars.

4. Theoretical Issues

Below, we have listed two prominent theoretical issues on axion stars. More details can be found in [2].

4.1. Emission from Axion Stars

As the axion is a real scalar field, the number of axions is not conserved. Self-interactions can convert the nonrelativistic axions into relativisitic ones. The axion stars and any other localized axion configuration with nonrelativisitic axions, inevitably radiate axion waves with relativistic wavelengths. As a result, the axion stars have finite lifetimes. It is important to calculate the lifetime of axion stars, since it determines whether they can have any observational significance.

NREFT appears to give unambiguous predictions for the conversion rate of nonrelativistic axions in axion stars into outgoing relativisitic axion waves [15]. The rate of decrease in N, the total number of nonrelativistic axions, is described by the anti-Hermitian terms in the effective Hamiltonian. When the number density of axions is small, such as in a dilute axion star, the loss of nonrelativistic axions is dominated by the decay into two photons. Thus the decay rate of the dilute axion star is the same as the decay rate of a single axion. The lifetime of a dilute axion star is therefore much longer than the age of the universe. For dense configurations, we define the lifetime to be the time required for the total number of axions to decrease by a factor $1/e$, when it moves to the left along the lower branch in Figure 2. In a dense axion star, the loss rate from the $4 \to 2$ process is approximately 5 orders of magnitude larger than that from $a \to 2\gamma$. The resulting predictions for the lifetime of the dense axion stars are still much longer than the age of the universe [15].

The predictions of NREFT for the loss rate of nonrelativistic axions are, however, incomplete. Surprisingly, there are loss processes for axions in the relativistic theory that cannot be reproduced by NREFT. NREFT is expect to correctly reproduce results from the corresponding relativistic theory for an oscillon with a small boson binding energy $\varepsilon_b \ll m_a$ as an expansion in powers of ε_b/m_a. However, such an expansion is blind to terms which are exponentially small in m_a/ε_b, such as $\exp(-c\sqrt{m_b/\varepsilon_b})$, where c is some constant. Thus, we should not expect a contribution having such an exponential factor to be reproduced by NREFT.

The contribution of loss processes whose rates have exponentially small factors can be calculated from the asymptotic expansion for the oscillon [25]. These terms have a radiative tail in the form of standing waves with exponentially small amplitudes that extend to infinity and have infinite energy. Without incoming waves, the outgoing waves decrease the total number of nonrelativistic axions of the localized part of the solution. The rate of decrease in the particle number, or equivalently the mass M, of the oscillon with angular frequency $\omega = \sqrt{1-\epsilon^2}\, m_a$ in the limit $\epsilon \to 0$ has the form [26]:

$$-\frac{dM}{dt} = \frac{A}{\epsilon^2} \exp(-3.406/\epsilon)\, f_a^2, \tag{15}$$

where the prefactor A depends on the axion potential $V(\phi)$. The sine-Gordon model is a special case in which A is suppressed by ϵ^2. For the sine-Gordon model in 3D, A is calculated to be $760.5\,\epsilon^2$ [26].

Eby et al. derived an expression for the loss rate that can be expressed in terms of the complex field $\psi(x)$ of NREFT [27]. Their derivation involves the matrix element of $V(\phi)$ inserted between an initial state of N condensed axions, each with energy $\omega = m_a - \varepsilon_b$, and a final state consisting of $N-3$ condensed axions plus an on-shell relativistic axion with energy 3ω. This can be interpreted as a $3 \to 1$ interaction, which is forbidden in the vacuum by conservation of momentum and energy. Their result for the rate of energy loss [27] can be expressed in the form:

$$-\frac{dM}{dt} = \frac{m_a \omega k}{192\pi f_a^4} \left| \int d^3 r\, e^{i\mathbf{k}\cdot\mathbf{r}} \left[\lambda_4 + \frac{\lambda_6 \psi^* \psi}{8 m_a f_a^2} + \ldots \right] \psi^3 \right|^2, \tag{16}$$

where $k^2 = 9\omega^2 - m_a^2$ and $\psi(r)$ is the wavefunction of the condensed axions normalized so the number of axions is $\int d^3 r\, \psi^* \psi$. A result consistent with Equation (16) was also obtained in [28], where this loss mechanism was referred to as "decay via spatial gradients". Since $|k| \approx \sqrt{8}\, m_a$, the loss comes from the small high-momentum tail of the wavefunction. For the instanton potential, its expansion in powers of $\psi^* \psi$ in Equation (16) can be summed up to all orders in terms of a Bessel function [27]. Eby et al.

obtained a result for the loss rate in Equation (16) for the sine-Gordon model in the limit $\varepsilon_b \to 0$ [29]. Their exponential suppression factor is consistent with Equation (15), but with the argument differing by less than 2%. Moreover, their result for the coefficient in the prefactor is $A = 2723$. It is different from the result in [26] by lacking a suppression factor of ϵ^2.

4.2. Collapse of Dilute Axion Stars

If a dilute axion star is embedded in a gas of unbounded axions, thermalization can condense additional axions and increase the mass of the axion star. For dilute axion stars close to the critical mass $M_* \approx N_* m_a$, where N_* is the total number of axions in the star, further condensation of axions can increase M to above M_*. Then, it will be unstable and collapse. The remnant of a collapsing dilute axion star has not been understood definitely. The possibilities for the remnant after the collapse include:

- A *black hole*, with a Schwarzschild radius which is smaller than the critical radius R_{99*} by about 15 orders of magnitude;
- A *dense axion star*, with a radius which is smaller than R_{99*} by about 7 orders of magnitude;
- A *dilute axion star*, with a radius which is larger than R_{99*}; and
- *No remnant*, because of complete disappearance into scalar waves.

Chavanis considered the possibility that a collapsing dilute axion star produces a black hole in [30]. The evolution of the axion field is obtained by solving the GPP equations for ψ and Φ given by Equation (13) with the truncated effective potential V_{eff} in Equation (14). He assumes the configuration for the complex axion field $\psi(r,t)$ can always be described by a Gaussian function, with a time-dependent radius $R(t)$. He found the time for collapse to $R = 0$ scales as $(M - M_*)^{-1/4}$ if the initial configuration is an unstable solution with mass M near M_*. Same variational methods were used previously to study the time evolution of gravitationally bound BECs of bosons with a positive scattering length [31].

Eby et al. also studied the collapse of dilute axion stars using a similar time-dependent variational approximation, but with V_{eff} given by the naive instanton effective potential in Equation (12) [32,33]. They found that the collapsing process is hindered by repulsive terms in the effective potential, which becomes important when the radius is close to that of a dense axion star. A large fraction of the total number of axions is lost through the emission of relativistic axions. But they were unable to determine definitely whether the remnant is a dense axion star.

Helfer et al. studied the fate of spherically symmetric axion configurations by solving the full nonlinear classical field equations in the framework of general relativity for axions with the instanton potential [34]. After evolving the configurations in time, they found the remnant could be a black hole or a dilute axion star or that there could be no remnant. Their calculations were limited to the parameter region $4 \times 10^{-8} < Gf_a^2 < 4 \times 10^{-2}$ and $0.03 < GMm_a < 0.12$. The three different possibilities for the remnant depend on different regions of the plane of Gf_a^2 versus GMm_a. The three regions meet at a triple point given by $Gf_a^2 = 3.6 \times 10^{-3}$ and $GMm_a = 0.095$. By extrapolating the results of [34] to the tiny value of Gf_a^2 for the QCD axion, one finds that the possibilities could be a black hole or no remnant.

Levkov, Panin, and Tkachev numerically calculated the collapse of dilute axion stars above the critical mass with the GPP equations [35]. Their solutions approach a self-similar scaling limit with a series of singularities at finite times t_*. Their calculation shows multiple cycles of growth of the energy density close to the center of the star followed by collapsing. The collapse dramatically increases the energy density near the center, followed by a burst of outgoing relativistic axion waves, which effectively depletes the energy density near the center. Levkov et al. also found that after these multiple cycles, the remnant is still gravitationally bound. They therefore concluded that the remnant must ultimately relax to a less-massive dilute axion star by gravitational cooling.

Funding: The research was supported in part by the DFG Collaborative Research Center "Neutrinos and Dark Matter in Astro- and Particle Physics" (SFB 1258).

Conflicts of Interest: The author declares no conflict of interest.

References

1. Kim, J.E.; Carosi, G. Axions and the strong CP problem. *Rev. Mod. Phys.* **2010**, *82*, 557. [CrossRef]
2. Braaten, E.; Zhang, H. Colloquium: The physics of axion stars. *Rev. Mod. Phys.* **2019**, *91*, 41002. [CrossRef]
3. Arvanitaki, A.; Dimopoulos, S.; Dubovsky, S.; Kaloper, N.; March-Russell, J. String axiverse. *Phys. Rev. D* **2010**, *81*, 123530. [CrossRef]
4. Hu, W.; Barkana, R.; Gruzinov, A. Fuzzy cold dark matter: the wave properties of ultralight particles. *Phys. Rev. Lett.* **2000**, *85*, 1158. [CrossRef] [PubMed]
5. Hui, L.; Ostriker, J.P.; Tremaine, S.; Witten, E. Ultralight scalars as cosmological dark matter. *Phys. Rev. D* **2017**, *95*, 43541. [CrossRef]
6. Peccei, R.D.; Quinn, H.R. CP Conservation in the Presence of Instantons. *Phys. Rev. Lett.* **1977**, *38*, 1440. [CrossRef]
7. Weinberg, S. A new light boson? *Phys. Rev. Lett.* **1978**, *40*, 223. [CrossRef]
8. Wilczek, F. Problem of strong P and T invariance in the presence of instantons. *Phys. Rev. Lett.* **1978**, *40*, 279. [CrossRef]
9. di Cortona, G.G.; Hardy, E.; Vega, J.P.; Villadoro, G. The QCD axion, precisely. *JHEP* **2016**, *1601*, 34. [CrossRef]
10. di Vecchia, P.; Veneziano, G.; Chiral dynamics in the large N limit. *Nucl. Phys. B* **1980**, *171*, 253. [CrossRef]
11. Bardeen, W.A.; Tye, S.-H.H.; Current algebra applied to properties of the light Higgs boson. *Phys. Lett. B* **1978**, *74*, 229. [CrossRef]
12. Peccei, R.D.; Quinn, H.R. Constraints imposed by CP conservation in the presence of pseudoparticles. *Phys. Rev. D* **1977**, *16*, 1791. [CrossRef]
13. Namjoo, M.H.; Guth, A.H.; Kaiser, D.I. Relativistic corrections to nonrelativistic effective field theories. *Phys. Rev. D* **2018**, *98*, 16011. [CrossRef]
14. Braaten, E.; Mohapatra, A.; Zhang, H. Nonrelativistic effective field theory for axions. Phys. Rev. D **2016**, *94*, 76004. [CrossRef]
15. Braaten, E.; Mohapatra, A.; Zhang, H. Emission of photons and relativistic axions from axion stars. *Phys. Rev. D* **2017**, *96*, 31901. [CrossRef]
16. Braaten, E.; Mohapatra, A.; Zhang, H. Classical nonrelativistic effective field theories for a real scalar field. *Phys. Rev. D* **2018**, *98*, 96012. [CrossRef]
17. Tkachev, I.I. On the possibility of Bose star formation. *Phys. Lett. B* **1991**, *261*, 289. [CrossRef]
18. Chavanis, P.H. Mass-radius relation of Newtonian self-gravitating Bose-Einstein condensates with short-range interactions: I. Analytical results. *Phys. Rev. D* **2011**, *84*, 43531. [CrossRef]
19. Schiappacasse, E.D.; Hertzberg, M.P. Analysis of dark matter axion clumps with spherical symmetry. *JCAP* **2018**, *1801*, 37. Erratum in **2018**, *1803*, E01. [CrossRef]
20. Eby, J.; Leembruggen, M.; Street, L.; Suranyi, P.; Wijewardhana, L.C.R. Approximation methods in the study of boson stars. *Phys. Rev. D* **2018**, *98*, 123013. [CrossRef]
21. Kling, F.; Rajaraman, A. Profiles of boson stars with self-interactions. *Phys. Rev. D* **2018**, *97*, 63012. [CrossRef]
22. Braaten, E.; Mohapatra, A.; Zhang, H. Dense Axion Stars. *Phys. Rev. Lett.* **2016**, *117*, 121801. [CrossRef] [PubMed]
23. Wang, X.Z. Cold Bose stars: Self-gravitating Bose-Einstein condensates. *Phys. Rev. D* **2001**, *64*, 124009. [CrossRef]
24. Visinelli, L.; Baum, S.; Redondo, J.; Freese, K.; Wilczek, F. Dilute and dense axion stars. *Phys. Lett. B* **2018**, *777*, 64. [CrossRef]
25. Segur, H.; Kruskal, M.D. Nonexistence of small amplitude breather solutions in ϕ^4 theory. *Phys. Rev. Lett.* **1987**, *58*, 747. [CrossRef] [PubMed]
26. Fodor, G.; Forgacs, P.; Horvath, Z.; Mezei, M. Radiation of scalar oscillons in 2 and 3 dimensions. *Phys. Lett. B* **2009**, *674*, 319. [CrossRef]
27. Eby, J.; Suranyi, P.; Wijewardhana, L.C.R. The lifetime of axion stars. *Mod. Phys. Lett. A* **2016**, *31*, 1650090. [CrossRef]

28. Mukaida, K.; Takimoto, M.; Yamada, M. On longevity of I-ball/oscillon. *JHEP* **2017**, *1703*, 122. [CrossRef]
29. Eby, J.; Ma, M.; Suranyi, P.; Wijewardhana, L.C.R. Decay of ultralight axion condensates. *JHEP* **2018**, *1801*. [CrossRef]
30. Chavanis, P.H. Collapse of a self-gravitating Bose-Einstein condensate with attractive self-interaction. *Phys. Rev. D* **2016**, *94*, 83007. [CrossRef]
31. Harko, T. Gravitational collapse of Bose-Einstein condensate dark matter halos. *Phys. Rev. D* **2014**, *89*, 84040. [CrossRef]
32. Eby, J.; Leembruggen, M.; Suranyi, P.; Wijewardhana, L.C.R. Collapse of axion stars. *JHEP* **2016**, *1612*. [CrossRef]
33. Eby, J.; Leembruggen, M.; Suranyi, P.; Wijewardhana, L.C.R. QCD axion star collapse with the chiral potential. *JHEP* **2017**, *1706*, 14. [CrossRef]
34. Helfer, T.; Marsh, D.J.E.; Clough, K.; Fairbairn, M.; Lim, E.A.; Becerril, R. Black hole formation from axion stars. *JCAP* **2017**, *1703*, 55. [CrossRef]
35. Levkov, D.G.; Panin, A.G.; Tkachev, I.I. Relativistic axions from collapsing Bose stars. *Phys. Rev. Lett.* **2017**, *118*, 11301. [CrossRef] [PubMed]

© 2019 by the author. Licensee MDPI, Basel, Switzerland. This article is an open access article distributed under the terms and conditions of the Creative Commons Attribution (CC BY) license (http://creativecommons.org/licenses/by/4.0/).

Article

The Grimus–Neufeld Model with FlexibleSUSY at One-Loop

Simonas Draukšas, Vytautas Dūdėnas, Thomas Gajdosik *, Andrius Juodagalvis, Paulius Juodsnukis and Darius Jurčiukonis

Institute of Theoretical Physics and Astronomy, Faculty of Physics, Vilnius University, Saulėtekio av. 3, 10257 Vilnius, Lithuania; s.drauksas@gmail.com (S.D.); vytautasdudenas@inbox.lt (V.D.); andrius.juodagalvis@tfai.vu.lt (A.J.); paulius.juodsnukis@gmail.com (P.J.); darius.jurciukonis@tfai.vu.lt (D.J.)
* Correspondence: tgajdosik@yahoo.com; Tel.: +370-611-69-425

Received: 28 October 2019; Accepted: 12 November 2019; Published: 16 November 2019

Abstract: The Grimus–Neufeld model can explain the smallness of measured neutrino masses by extending the Standard Model with a single heavy neutrino and a second Higgs doublet, using the seesaw mechanism and radiative mass generation. The Grimus–Lavoura approximation allows us to calculate the light neutrino masses analytically. By inverting these analytic expressions, we determine the neutrino Yukawa couplings from the measured neutrino mass differences and the neutrino mixing matrix. Short-cutting the full renormalization of the model, we implement the Grimus–Neufeld model in the spectrum calculator FlexibleSUSY and check the consistency of the implementation. These checks hint that FlexibleSUSY is able to do the job of numerical renormalization in a restricted parameter space. As a summary, we also comment on further steps of the implementation and the use of FlexibleSUSY for the model.

Keywords: neutrinos; seesaw mechanism; radiative masses; spectrum calculator

PACS: 11.30.Rd; 13.15.+g; 14.60.St

1. Introduction

The last 30 years of collider physics showed an ever-increasing success for the predictions of the Standard Model (SM) [1]. A similar statement can be said about the experimental program in neutrino physics [2], but no unambiguous common treatment for both areas exists up to now. The masses and the mixing of neutrinos can easily be formulated in a Lagrangian picture; nevertheless, these terms in the Lagrangian are still considered to be "beyond the Standard Model" (BSM). Whereas the formulation of the SM and the accurate calculations for the experimental predictions require the framework of Quantum Feld Theory (QFT), the analysis of neutrino measurements is still done in the framework of plain Quantum Mechanics (QM); an explanation of why QM is usually enough for the study of neutrino oscillations can be found in [3].

The usual two explanations of the smallness of neutrino masses in the Lagrangian context are the seesaw mechanism [4] or radiative mass generation for the light neutrinos [5–8]. In 1989, Walter Grimus and Helmut Neufeld pointed to the possibility that both mechanisms can be comparable for explaining the masses of the light neutrinos [9]. We call the minimal extension of the SM that allows this feature Grimus–Neufeld Model (GNM). This minimal extension adds to the SM only a single heavier Majorana singlet and one additional Higgs doublet. The qualitative behavior of the GNM is described in [10]. An extended description of our approach is presented in [11]. Here, we review shortly the features of the GNM and discuss its implementation in FlexibleSUSY [12,13], which uses SARAH [14–17] and SOFTSUSY [18,19]. This implementation gives us a tool to check the renormalization and the consistency of the model numerically.

The main aim of this work is to discuss the plans for the checks of the model and the possible modifications that can be done to the code. The first trials of the implemented code hint that, to have only small loop corrections, the model has a natural preference for a small seesaw scale.

2. Results

2.1. Summary of Features of the Grimus–Neufeld Model

2.1.1. Lagrangian

Since the GNM is a minimal extension of the SM, we only need to give the additional parts of the Lagrangian. These are the Majorana mass term for the fermionic singlet N

$$-\mathcal{L}_{\text{Majorana-mass}} = \tfrac{1}{2}\bar{N}_R M_R \hat{N}_R + h.c. = -\tfrac{1}{2}\bar{N}_R M_R C \bar{N}_R^\top + h.c. \, , \tag{1}$$

the Yukawa terms (ignoring quarks) among the lepton doublets ℓ_j, the Higgs doublets Φ_a, and either the charged lepton singlets E_k or the fermionic singlet N

$$-\mathcal{L}_{\text{F-H}} = \bar{\ell}_j \Phi_a (Y_E^{(a)})_{jk} E_k + \bar{\ell}_j \tilde{\Phi}_a (Y_N^{(a)})_j N + h.c. \, , \tag{2}$$

and the Higgs potential

$$-\mathcal{L}_{\text{H}} = V(\Phi_a) = Y_{ab}(\Phi_a^\dagger \Phi_b) + \tfrac{1}{2} Z_{abcd}(\Phi_a^\dagger \Phi_b)(\Phi_c^\dagger \Phi_d) \, , \tag{3}$$

which is just the generic two Higgs doublet potential, which we write in the Higgs basis [20], meaning that only Φ_1 has a vacuum expectation value (vev) v. $\hat{N}_R$ in Equation (1) is the Lorentz covariant conjugate [21] of $N_R = P_R N$, which is the right-chiral part of N.

2.1.2. Tree-Level Mass Matrices and Tree-Level Masses

As we are interested mainly in the neutrino sector, we deal only with the mass matrices of the leptons and the Higgs bosons. For the charged leptons, the relevant Yukawa coupling is $Y_E^{(1)}$, since we work in the Higgs basis. Then, diagonalization of the charged lepton mass matrix M_E,

$$U_{eL} M_E U_{eR}^\dagger = U_{eL}(\tfrac{v}{\sqrt{2}} Y_E^{(1)}) U_{eR}^\dagger = \text{diag}(m_e, m_\mu, m_\tau) \, , \tag{4}$$

defines the mass eigenstates for the charged leptons and, consequently, the flavor basis for the SM neutrino fields $(\nu_e, \nu_\mu, \nu_\tau)$, which are the partners of the charged leptons in the weak lepton doublets.

The mass matrix for the neutrinos in the GNM in flavor basis at tree level is the symmetric 4×4-matrix M_ν,

$$M_\nu = \begin{pmatrix} M_L & M_D \\ M_D^\top & M_R \end{pmatrix} \, , \quad \text{with} \quad \begin{aligned} M_L &= 0_{3\times 3} \quad \text{at tree level} \\ M_D &= \tfrac{v}{\sqrt{2}} Y_N^{(1)} \end{aligned} \, , \tag{5}$$

which is only rank 2 and therefore has only two non-vanishing singular values. That means we have the heavy mass $m_4^{(0)} \sim M_R$ and only a single light, massive neutrino with mass $m_s^{(0)}$ coming from the seesaw. The two other masses $m_o^{(0)} = m_r^{(0)} = 0$ are zero at tree level. The radiative mass will be $m_r > 0$ at one-loop, but the remaining other neutrino with mass $m_o = 0$ stays massless even at one-loop level.

Following the idea in [22,23] to formulate the 2HDM potential in terms of basis independent physical quantities, we skip the discussion of the mass matrix of the Higgs bosons and point the reader to the relevant literature [20,24–26]. For the tree-level masses of the Higgs bosons, we just want to note that we take the lightest boson, h, to correspond to the boson observed at the LHC with the mass $m_h = 125.18\,\text{GeV}$ [1]. The other Higgs boson masses are free parameters subject to the experimental constraints.

2.1.3. Leading Order Loop-Level Masses

The observation that the predicted loop-level mass for one neutrino, m_r, can be of the same order as the seesaw generated mass $m_s^{(0)}$ was the main point of the paper of Grimus and Neufeld [9] in 1989. Since the remaining neutrino stays massless at one-loop order, $m_o^{(1)} = 0$, and all other particles already have a mass (unless they are protected by an unbroken gauge symmetry), this radiative neutrino mass is the main effect at one loop. This predicted mass m_r is finite and gauge invariant, as proven in [10,27,28] using different approaches. One loop radiative corrections affect the seesaw generated neutrino mass $m_s^{(0)}$, too. We denote the resulting mass as m_s.

2.2. The Grimus–Lavoura Approximation

The full calculation of the renormalized neutrino masses for the GNM in analytic form is not available yet. We adapt the proof of finiteness and gauge invariance of the one-loop corrections to the effective light neutrino mass matrix from [27] and formulate the Grimus–Lavoura approximation for calculating light neutrino masses:

- Staying in the interaction eigenstates, as defined by the charged leptons, calculate the neutrino mass matrix M_ν using Equation (5).
- The light neutrino block M_L does not have a counter-term $\delta^{ct} M_L = 0$ at one loop level, as $M_L^{(0)}$ vanishes at tree-level. The loop corrections δM_L, depicted by the diagrams in Figure 1, are finite and gauge invariant, as proven in [27].
- Reducing the problem to the light neutrinos, one arrives at the effective symmetric 3×3 neutrino mass matrix $\mathcal{M}_\nu$, which has the tree-level value

$$\mathcal{M}_\nu^{\text{tree}} = -M_D M_R^{-1} M_D^\top , \qquad (6)$$

that has obviously rank 1 and will give only the neutrino mass $m_s^{(0)}$. However, at one-loop, this matrix becomes

$$\mathcal{M}_\nu^{\text{1-loop}} = \mathcal{M}_\nu^{\text{tree}} + \delta \mathcal{M}_\nu , \qquad (7)$$

with

$$\delta \mathcal{M}_\nu = \delta M_L - \delta M_D M_R^{-1} M_D^\top - M_D M_R^{-1} \delta M_D^\top - M_D M_R^{-1} (\delta M_R) M_R^{-1} M_D^\top . \qquad (8)$$

With this correction, $\mathcal{M}_\nu^{\text{1-loop}}$ can have rank > 1.
- The approximation consists now of:

  1. Assuming δM_R to be irrelevant for the light neutrinos with the reasoning that M_R (or m_4) is not measured. It is still a free parameter of the theory that can be adjusted as needed.
  2. Observing that the corrections with δM_D are subdominant, because they are suppressed by the squares of small Yukawa or gauge couplings and additionally by the small charged lepton masses.
  3. Assuming that the loop correction δM_L is of the same order as the tree-level value $\mathcal{M}_\nu^{\text{tree}}$.

The result of this approximation is that we can derive analytic formulas that predict the masses of the light neutrinos, as depending on the tree-level input parameters:

$$m_{r,s} = m_{r,s}(m_{H_k^0}^2, \vartheta_{1j}; m_4^{(0)}, Y_N^{(a)}) , \qquad (9)$$

where $m_4^{(0)} \sim m_4 \sim M_R$. ϑ_{1j} are the basis-independent mixing angles of the neutral Higgs fields [20]. Following Grzadkowski et al. [22], these angles can be expressed by the $H_j^0 W^+ W^-$-couplings $\frac{1}{2}g^2 e_j$ as

$$e_1 = v c_{12} c_{13}, \quad e_2 = v s_{12} c_{13}, \quad e_3 = v s_{13}, \quad \text{where} \quad c_{1j} = \cos \vartheta_{1j} \text{ and } s_{1j} = \sin \vartheta_{1j}. \tag{10}$$

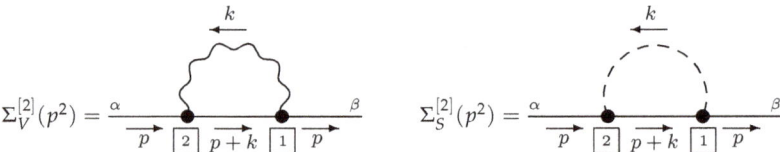

Figure 1. Feynman diagrams contributing to the self-energies of the light neutrinos. For the correction to the mass, the internal fermion line has to be a Majorana propagator with a mass insertion; hence, charged particles will not contribute to δM_L at this loop level.

2.3. Using the Grimus–Lavoura Approximation

Since our model predicts one neutrino mass to remain zero at one-loop level ($m_o^{(0)} = m_o^{(1)} = 0$), we can use the measured neutrino mass squared differences [2] to determine the values of the other light neutrino masses $m_{r,s}$.

Parameterizing the neutrino Yukawa couplings as

$$(Y_N^{(1)})_k = \frac{\sqrt{2} m_D}{v} u_{3k} \quad \text{and} \quad (Y_N^{(2)})_k := d\, u_{2k} + d'\, u_{3k}, \tag{11}$$

with three orthonormal three-vectors $u_{\alpha k} = (\vec{u}_\alpha)_k$, we can invert the analytic expressions of the masses, Equation (9), and determine the parameters d and $|d'|$. The explicit formulas and the discussion of the difficulties in finding solutions for $|d'|$ can be found in [11]. The values of d and $|d'|$ do not depend on the three orthonormal three-vectors $u_{\alpha k}$. At tree-level, these vectors $u_{\alpha k}$ can be understood as an approximate neutrino mixing matrix $\tilde{V}_{\text{PMNS}}$, as it diagonalizes the effective tree-level neutrino mass matrix with the Takagi decomposition:

$$u_{\alpha j}^* (\mathcal{M}_\nu^{\text{tree}})_{jk} u_{\beta k}^* = (\tilde{V}_{\text{PMNS}}^\top)_{\alpha j} (\mathcal{M}_\nu^{\text{tree}})_{jk} (\tilde{V}_{\text{PMNS}})_{k\beta} = -\text{diag}(m_o^{(0)} = 0, m_r^{(0)} = 0, m_s^{(0)} > 0). \tag{12}$$

Applying these same vectors to the effective one-loop neutrino mass matrix

$$u_{\alpha j}^* (\mathcal{M}_\nu^{\text{1-loop}})_{jk} u_{\beta k}^* = \begin{pmatrix} 0 & 0 & 0 \\ 0 & A & B \\ 0 & B & C \end{pmatrix}_{\alpha \beta}, \tag{13}$$

we see that the effective one-loop neutrino mass matrix is rank 2, and hence provides two massive light neutrino states. The full neutrino mixing matrix V_{PMNS} should diagonalize the full one-loop mass matrix

$$(V_{\text{PMNS}}^\top)_{\alpha j} (\mathcal{M}_\nu^{\text{1-loop}})_{jk} (V_{\text{PMNS}})_{k\beta} = \text{diag}(m_o^{(1)} = 0, m_r^{(1)} > 0, m_s^{(1)} > 0). \tag{14}$$

A reordering of the masses is possible, if $m_r^{(1)} > m_s^{(1)}$. For a more detailed discussion, see [11].

As a side note: The minus-sign on the right-hand side of Equation (12) comes from the convention of the seesaw, where the seesaw rotation is written with an orthogonal matrix and the minus sign kept to be absorbed in the phase of the light state. We do not write a minus-sign on the right-hand side of Equation (14) following the convention of the normal singular value decomposition and the positivity of the masses. The phase in this second case is part of the complex mixing matrix.

Diagonalizing the reduced effective one-loop neutrino mass matrix, Equation (13), we get an effective 2×2 mixing matrix that connects the full mixing matrix V_{PMNS} with the used vectors $u_{\alpha k}$. This finally allows us to determine the vectors $u_{\alpha k}$, Equation (11), from the measured PMNS matrix:

$$u_{\alpha k} = u_{\alpha k}(m_4, \Delta m_{\text{atm}}^2, \Delta m_{\text{sol}}^2, V_{\text{PMNS}}; m_{H_k^0}^2, \vartheta_{1j}; \tfrac{\sqrt{2}m_D}{v}, \phi' = \arg[d']) \ . \tag{15}$$

2.4. The One-Loop Improved Lagrangian

Using the neutrino Yukawa couplings as defined in the previous subsection, the determination of the Higgs potential parameters following Grzadkowski et al. [22], and the identification of the singlet mass parameter with the heavy neutrino mass $M_R \sim m_4$, we get a new parametrization of the Lagrangian as

$$-\mathcal{L}'_{\text{Majorana-mass}} = \tfrac{1}{2}\bar{N}_R m_4 \widehat{N}_R + h.c. = -\tfrac{1}{2}\bar{N}_R m_4 \mathbf{C} \bar{N}_R^\top + h.c. \ , \tag{16}$$

the Yukawa terms

$$-\mathcal{L}'_{\text{F-H}} = \bar{\ell}_j \Phi_a (Y_E^{(a)})_{jk} E_k + (\tfrac{\sqrt{2}m_D}{v}\tilde{\Phi}_1 + d'\tilde{\Phi}_2) u_{3j} \bar{\ell}_j P_R N + d\,\tilde{\Phi}_2 u_{2j} \bar{\ell}_j P_R N + h.c. \ , \tag{17}$$

where d and d' are functions depending on the same parameters as $u_{\alpha j}$, Equation (15), and the Higgs potential

$$-\mathcal{L}'_{\text{H}} = V(\Phi_a; Y, Z) \quad \text{with } \{Y, Z\} = \{Y, Z\}(m_{H^\pm}^2, m_{H_k^0}^2, e_j, q_j, q) \ , \tag{18}$$

is expressed in terms of physical masses and couplings of the Higgs bosons.

As an advantage, more parameters of the one-loop improved Lagrangian $\mathcal{L}'$ correspond directly to measured quantities: instead of six complex parameters in two-neutrino Yukawa couplings in $\mathcal{L}$, $\mathcal{L}'$ contains two real selectable parameters ($\tfrac{\sqrt{2}m_D}{v}$ and ϕ'), two real parameters that we determine from the measured neutrino mass squared differences (d and $|d'|$), and six parameters in the vectors $u_{\alpha j}$ that we determine from the measured neutrino mixing matrix V_{PMNS}. One of the two "missing" parameters in $\mathcal{L}'$ is the vanishing one-loop level neutrino mass, and the other parameter is the undeterminable Majorana phase of this zero-mass neutrino.

2.5. Renormalizing the GNM

The final goal of our efforts is to fully renormalize the GNM. The full renormalization will also indicate the importance and validity of the one-loop improved Lagrangian. However, we are still far from that goal. Only the mass renormalization of light neutrinos has been tackled in detail [28,29]. Since the full renormalization is difficult, we plan to use a spectrum calculator that performs the renormalization numerically. A foreseeable difficulty lies in the hierarchy of the seesaw. It is hard to have a reliable numerical implementation of the mass hierarchies of more than 10 orders of magnitude. We found that FlexibleSUSY [13] is able to do the job if we limit the seesaw scale. Now, we can study the renormalized model, but only numerically.

2.6. FlexibleSUSY for the GNM in a Nutshell

The primary idea of FlexibleSUSY was to numerically implement the renormalization group running of a model between two scales and to be able to give boundary conditions on both scales. In the case of a SUSY-GUT, one can require the low energy measured masses and couplings as one boundary condition and the GUT constraints at the GUT scale as another boundary condition. Then, the program tries to find a numerical solution that interpolates between the two scales and fulfills both boundary conditions.

In our case, we do not have the high scale, and we are not interested in imposing conditions at a scale other than our low scale. However, with the accuracy needed for today's collider physics, one has to interpolate between the various scales of the different measurements that determine the masses and couplings of the SM and also the masses and mixing parameters of the neutrino sector. That means that the capabilities of FlexibleSUSY are needed also for implementing models that "live" only at the low scale.

To implement a model in FlexibleSUSY, one has first to define the Lagrangian of the model with SARAH [14–17] and check the consistency and completeness of the Lagrangian. In the next step, FlexibleSUSY uses SARAH to produce the model code. Additionally, one has to define the boundary conditions in a separate steering file, using the convention for the names of fields implemented in SARAH. This step includes the definition of what is to be treated as input and what is the desired output for the spectrum to be calculated. When this is done, one has to compile the generated code to get the actual spectrum calculator for the implemented model.

This program can be used from the terminal with the help of input and output files in the SLHA [30,31] format. That means writing the values of the input parameters and additional optional arguments into the SLHA input file and the program writes from this input the SLHA output file, which contains the mass spectrum of the model and also the decay rates of the particles of the model.

Another option to use the program is from the provided Mathematica^{TM} [32] interface. There, the playing around with parameters and the generation of plots become much easier, but the comparison of results with other scientists becomes less accurate or much more cumbersome.

2.6.1. Our Achievements with the GNM in FlexibleSUSY

We succeeded to implement the basic Lagrangian of the GNM, $\mathcal{L}$ in Equations (1)–(3), in FlexibleSUSY and to generate a working code. This code could qualitatively reproduce the effect of the seesaw mechanism. However, it was difficult to find a point in the Higgs potential that gives sensible results for the Higgs masses also at one-loop level. This might be due to our limited experience with spectrum calculators and numerics. However, we are learning a lot of physics while trying to figure out where the problems can come from.

2.6.2. Plans with the GNM in FlexibleSUSY

The first step is to check our implementation:

- Do we understand the tree-level correctly?
- Are the different formulations of defining the input parameters really equivalent?
- Does the GL approximation give correct neutrino masses not only at leading order, but also at full one-loop level?
- Is the limitation of the seesaw scale a numeric artifact from the finite precision or can we find a physical reason for the limitation?

The next step is to investigate the parameter space of the GNM at one-loop level, i.e., going beyond the analysis of Jurčiukonis et al. [11]. The real restrictions to the model come from comparing to measurements. We hope to recycle the FlexibleSUSY implementations of predictions of other models that are already implemented in FlexibleSUSY. One definite goal is to work out the connection between the low-energy observables such as $(g-2)_\mu$, $\mu \to 3e$, and $\mu \to e\gamma$, which is caused by the GNM. Another goal for the future is to work out the implications of the GNM for cosmology, where the question might be, if the heavy singlet can be a candidate for the dark matter.

3. Discussion

The main result is the implementation and the analytic check of the Grimus–Lavoura approximation in the Grimus–Neufeld model (GNM) to replace the Yukawa couplings with the measured masses and oscillation parameters as the input for the model. The model itself [9] and the approximation [33] do not try to estimate the model parameters from measurements. In addition, the renormalization group analysis of this model in the limit of heavy right-handed neutrinos and heavy Higgs doublets [10] does not provide the explicit analytical and numerical analysis that is given in [11] and is shortly recapitulated in this paper.

The additional content of this presentation compared to the one in [11] is to address the question of the full renormalization of the GNM. The first steps in that direction were done in [29], but the full and explicit formulation of the renormalization of the GNM is work in progress with its end still far away.

To shorten the time to some results, we propose the use of generic spectrum calculators and give an account on the progress achieved with FlexibleSUSY, which is such a spectrum calculator. The final goal from both approaches, of the analytic one with fully renormalizing the model and of the numeric one with FlexibleSUSY, is to test the GNM with measurements. To that end, we want to give predictions that can be tested, such as correlations of low energy observables or correlations between the decay rates of heavy particles, which are influenced by the neutrino Yukawa coupling.

Author Contributions: Conceptualization: T.G.; methodology: T.G., S.D. and V.D.; software: S.D. and P.J.; validation: T.G., V.D. and D.J.; formal analysis: D.J.; investigation: S.D., V.D., P.J. and D.J.; data curation: D.J.; writing—original draft preparation: T.G.; writing—review and editing: T.G., S.D., V.D., A.J., P.J. and D.J.; visualization: T.G.; supervision: T.G. and A.J.; project administration: T.G. and A.J.; and funding acquisition: A.J.

Funding: This research was funded by the Lithuanian Academy of Sciences.

Acknowledgments: The authors are thankful for the very helpful discussions with Dominik Stöckinger and Wojciech Kotlarski at the conference "Matter to the Deepest 2019".

Conflicts of Interest: The authors declare no conflict of interest.

Abbreviations

The following abbreviations are used in this manuscript:

2HDM	two Higgs doublet model
BSM	beyond the standard model physics
GNM	Grimus–Neufeld model
GUT	grand unified theory
LHC	large hadron collider
PMNS	Pontecorvo–Maki–Nakagawa–Sakata matrix
QFT	quantum field theory
QM	quantum mechanics
SM	standard model

References

1. Tanabashi, M.; Hagiwara, K.; Hikasa, K.; Nakamura, K.; Sumino, Y.; Takahashi, F.; Tanaka, J.; Agashe, K.; Aielli, G.; Amsler, C.; et al. Review of Particle Physics. *Phys. Rev. D* **2018**, *98*, 030001. [CrossRef]
2. De Salas, P.F.; Forero, D.V.; Ternes, C.A.; Tortola, M.; Valle, J.W.F. Status of neutrino oscillations 2018: 3σ hint for normal mass ordering and improved CP sensitivity. *Phys. Lett. B* **2018**, *782*, 633–640. [CrossRef]
3. Akhmedov, E.K.; Kopp, J. Neutrino Oscillations: Quantum Mechanics vs. Quantum Field Theory. *J. High Energy Phys.* **2010**. [CrossRef]
4. Schechter, J.; Valle, J. Neutrino Masses in SU(2) x U(1) Theories. *Phys. Rev. D* **1980**, *22*, 2227. [CrossRef]
5. Zee, A. A Theory of Lepton Number Violation, Neutrino Majorana Mass, and Oscillation. *Phys. Lett. B* **1980**, *93*, 389. [CrossRef]
6. Branco, G.C.; Geng, C.Q. Naturally Small Dirac Neutrino Masses in Superstring Theories. *Phys. Rev. Lett.* **1987**, *58*, 969. [CrossRef]
7. Chang, D.; Mohapatra, R.N. Small and Calculable Dirac Neutrino Mass. *Phys. Rev. Lett.* **1987**, *58*, 1600. [CrossRef]
8. Babu, K.S. Model of 'Calculable' Majorana Neutrino Masses. *Phys. Lett. B* **1988**, *203*, 132–136. [CrossRef]
9. Grimus, W.; Neufeld, H. Radiative Neutrino Masses in an SU(2) X U(1) Model. *Nucl. Phys. B* **1989**, *325*, 18. [CrossRef]

10. Ibarra, A.; Simonetto, C. Understanding neutrino properties from decoupling right-handed neutrinos and extra Higgs doublets. *J. High Energy Phys.* **2011**, *11*, 022. [CrossRef]
11. Jurčiukonis, D.; Gajdosik, T.; Juodagalvis, A. Seesaw neutrinos with one right-handed singlet field and a second Higgs doublet. *arXiv* **2019**, arXiv:hep-ph/1909.00752.
12. Athron, P.; Park, J.H.; Stöckinger, D.; Voigt, A. FlexibleSUSY—A spectrum generator generator for supersymmetric models. *Comput. Phys. Commun.* **2015**, *190*, 139–172. [CrossRef]
13. Athron, P.; Bach, M.; Harries, D.; Kwasnitza, T.; Park, J.H.; Stöckinger, D.; Voigt, A.; Ziebell, J. FlexibleSUSY 2.0: Extensions to investigate the phenomenology of SUSY and non-SUSY models. *arXiv* **2017**, arXiv:hep-ph/1710.03760.
14. Staub, F. From Superpotential to Model Files for FeynArts and CalcHep/CompHep. *Comput. Phys. Commun.* **2010**, *181*, 1077–1086. [CrossRef]
15. Staub, F. Automatic Calculation of supersymmetric Renormalization Group Equations and Self Energies. *Comput. Phys. Commun.* **2011**, *182*, 808–833. [CrossRef]
16. Staub, F. SARAH 3.2: Dirac Gauginos, UFO output, and more. *Comput. Phys. Commun.* **2013**, *184*, 1792–1809. [CrossRef]
17. Staub, F. SARAH 4: A tool for (not only SUSY) model builders. *Comput. Phys. Commun.* **2014**, *185*, 1773–1790. [CrossRef]
18. Allanach, B. SOFTSUSY: A program for calculating supersymmetric spectra. *Comput. Phys. Commun.* **2002**, *143*, 305–331. [CrossRef]
19. Allanach, B.; Athron, P.; Tunstall, L.C.; Voigt, A.; Williams, A. Next-to-Minimal SOFTSUSY. *Comput. Phys. Commun.* **2014**, *185*, 2322–2339. [CrossRef]
20. Haber, H.E.; O'Neil, D. Basis-independent methods for the two-Higgs-doublet model. II. The Significance of tan beta. *Phys. Rev. D* **2006**, *74*, 015018. [CrossRef]
21. Pal, P.B. Dirac, Majorana and Weyl fermions. *Am. J. Phys.* **2011**, *79*, 485–498. [CrossRef]
22. Grzadkowski, B.; Haber, H.E.; Ogreid, O.M.; Osland, P. Heavy Higgs boson decays in the alignment limit of the 2HDM. *J. High Energy Phys.* **2018**, *12*, 056. [CrossRef]
23. Grzadkowski, B.; Ogreid, O.M.; Osland, P. The CP-symmetries of the 2HDM. In Proceedings of the 6th Symposium on Prospects in the Physics of Discrete Symmetries (DISCRETE 2018), Vienna, Austria, 26–30 November 2018.
24. Davidson, S.; Haber, H.E. Basis-independent methods for the two-Higgs-doublet model. *Phys. Rev. D* **2005**, *72*, 035004. [CrossRef]
25. Haber, H.E.; O'Neil, D. Basis-independent methods for the two-Higgs-doublet model III: The CP-conserving limit, custodial symmetry, and the oblique parameters S, T, U. *Phys. Rev. D* **2011**, *83*, 055017. [CrossRef]
26. Branco, G.C.; Ferreira, P.M.; Lavoura, L.; Rebelo, M.N.; Sher, M.; Silva, J.P. Theory and phenomenology of two-Higgs-doublet models. *Phys. Rept.* **2012**, *516*, 1–102. [CrossRef]
27. Grimus, W.; Lavoura, L. Soft lepton flavor violation in a multi Higgs doublet seesaw model. *Phys. Rev. D* **2002**, *66*, 014016. [CrossRef]
28. Dūdėnas, V.; Gajdosik, T. Gauge dependence of tadpole and mass renormalization for a seesaw extended 2HDM. *Phys. Rev. D* **2018**, *98*, 035034. [CrossRef]
29. Dūdėnas, V. *Renormalization of Neutrino Masses in the Grimus-Neufeld Model*; Vilniaus Universiteto Leidykla: Vilnius, Lithuania, 2019.
30. Allanacha, B.C.; Balázs, C.; Bélanger, G.; Bernhardt, M.; Boudjema, F.; Choudhury, D.; Desch, K.; Ellwanger, U.; Gambino, P.; Godbole, R.; et al. SUSY Les Houches Accord 2. *Comput. Phys. Commun.* **2009**, *180*, 8–25. [CrossRef]
31. Basso, L.; Belyaev, A.; Chowdhury, D.; Hirsch, M.; Khalil, S.; Moretti, S.; O'Leary, B.; Porod, W.; Staub, F. Proposal for generalised Supersymmetry Les Houches Accord for see-saw models and PDG numbering scheme. *Comput. Phys. Commun.* **2013**, *184*, 698–719. [CrossRef]
32. Wolfram Research, Inc. *Mathematica*, Version 12.0; Wolfram Research, Inc.: Champaign, IL, USA, 2019.
33. Grimus, W.; Lavoura, L. One loop corrections to the seesaw mechanism in the multiHiggs doublet standard model. *Phys. Lett. B* **2002**, *546*, 86–95. [CrossRef]

© 2019 by the authors. Licensee MDPI, Basel, Switzerland. This article is an open access article distributed under the terms and conditions of the Creative Commons Attribution (CC BY) license (http://creativecommons.org/licenses/by/4.0/).

Article

Exponentiation in QED and Quasi-Stable Charged Particles

Stanisław Jadach [1], Wiesław Płaczek [2] and Maciej Skrzypek [1,*]

[1] Institute of Nuclear Physics Polish Academy of Sciences, ul. Radzikowskiego 152, 31-342 Kraków, Poland; stanislaw.jadach@ifj.edu.pl
[2] Marian Smoluchowski Institute of Physics, Jagiellonian University, ul. Łojasiewicza 11, 30-348 Kraków, Poland; wieslaw.placzek@uj.edu.pl
* Correspondence: maciej.skrzypek@ifj.edu.pl

Received: 16 October 2019; Accepted: 3 November 2019; Published: 8 November 2019

Abstract: In this note we present a new exponentiation scheme of soft photon radiation from charged quasi-stable resonances. It generalizes the well established scheme of Yennie, Frautschi and Suura. While keeping the same functional form of an exponent, the new scheme is both exact in its soft limit and accounts properly for the kinematical shift in resonant propagators. We present the scheme on an example of two processes: a toy model of single W production in $e\nu$ scattering and the W pair production and decay in e^+e^- annihilation. The latter process is of relevance for the planned FCCee collider where high precision of Monte Carlo simulations is a primary goal. The proposed scheme is a step in this direction.

Keywords: QED; exponentiation; W-boson

PACS: 12.20.-m; 14.70.Fm

1. Introduction

Emission of soft photons accompany every process with charged particles. Therefore proper treatment of such emission is mandatory. These emissions happen from the external particles, decouple from the process itself and can be resummed and exponentiated in an universal way [1]. In such an approach the information on the photonic emission is not transmitted to the process itself. This can pose a problem if the process includes resonant particles in the intermediate state, because any energy loss due to photonic emission, larger compared to the width of the resonance can shift the process off resonance and the Γ/M suppression should be visible. We will refer to it as the recoil effect. This effect for the case of neutral resonance has been resolved with the help of coherent states in References [2,3], and then at the level of spin amplitudes in Reference [4].

In the case of charged resonances another complication appears: the resonance is a source of the soft photons as well. Of course, strictly speaking, the internal particles do not radiate soft photons (do not have singularities due to such radiation), as demonstrated by Yennie, Frautschi and Suura (YFS61) in the classical Reference [1]. However, the resonances are special – they are quasi-stable and there is a clear separation between their production and decay. One can illustrate it with the case of lepton τ for which the lifetime is longer than the time-scale of the production by an astronomical factor of $m_\tau/\Gamma_\tau \simeq 3 \times 10^{11}$. Resummation (exponentiation) of such emissions is the subject of this note. We will present a solution that smoothly interpolates between two situations: for $\sum k^0 < \Gamma$ we have the normal YFS61 behaviour, i.e. internal radiation is suppressed, whereas for $\sum k^0 > \Gamma$ the recoil effect is properly accounted for. More details can be found in a recent paper [5].

We analyse two processes: the simplest toy model, $e\bar{\nu}_e \to W \to \mu\bar{\nu}_\mu$, on which we demonstrate the methodology; and with an eye on future e^+e^- collider, the full-scale process, $e^+e^- \to WW \to 4f$.

The latter process is one of a few gold-plated processes of the projected FCCee machine. In its second phase, at the WW-threshold, the FCCee would provide about 3×10^7 WW events. That number corresponds to a statistical error on the total cross section of 0.02%, or equivalently $\Delta M_W \simeq 0.3$ MeV (measured from the threshold scan). The current state of the art, inherited from the LEP2 era, is 0.5–2% for the total cross section. That means, an increase of the precision by factor of 100 is needed. The most important part of that challenge is to calculate the $\mathcal{O}(\alpha^2)$ corrections to the signal process $e^+e^- \to WW \to 4f$. Exponentiation of soft radiation from Ws, which is equivalent to resummation (exact in the soft limit) of photon interferences between production and decay stages as well as between decays of the two Ws would encapsulate an important part of these corrections, not only to the second order but to all orders!

There is also another, practical, and perhaps even most important, application of the soft YFS-based exponentiation—the Monte Carlo event generators. The whole family of such generators has emerged from the never published note of S. Jadach [6]. All of them generate multiple sof-photon emissions based on the classical YFS61 exponentiation. Among them one should list BHLUMI [7] and KORALZ [8]. The latter one has been then replaced by a next-generation-code KKMC [9] which includes recoil in production-decay interferences of the Z-resonance. The partial solutions related to the charged W bosons have been implemented in the YFSWW3 program [10] where exponentiation of the emission in the WW pair production ($ee \to WW$) with the recoil has been done and in WINHAC [11] in which exponentiation of radiation in the W decay ($W \to f_1 f_2$) is implemented for the single-W process.

At last, let us only touch upon the issue of QED deconvolution. The YFS approach provides for a very convenient scheme of such a deconvolution also in higher orders. When supplied with the treatment of resonances, it would form a complete and well defined deconvolution system.

The paper is organized as follows. In Section 2 we present the standard YFS scheme and its new extension on the example of the simple "toy model" $e\bar{v}_e \to W \to \mu\bar{v}_\mu$. In Section 3 we discuss the exponentiation in the process of W-pair production and decay. In particular we show how to introduce virtual soft-photon interferences and how to exponentiate them. The last section contains conclusions and summary.

2. Toy Model with Single W

In this section we show, in the combinatorial language, on the simplest possible example of $e(p_a)\bar{v}_e(p_b) \to W \to \mu(p_c)\bar{v}_\mu(p_d)$, how the YFS61 procedure works and then we extend it by adding the soft real emission from the semi-stable W boson. The combinatorial resummation presented here differs from the original YFS61 derivation of Ref. [1] which was done with the help of Bose symmetry principles.

2.1. Classical YFS Resummation from External Legs

The YFS61 theorem states that soft emissions are singular only from external legs. For the single emission in a generic process $\mathcal{M}$, depicted in Figure 1 we have

$$\mathcal{M}^\mu \simeq \mathcal{M}^{(0)}(p-k)\frac{\mathbf{p}-\mathbf{k}+m}{k^2-2kp+i\varepsilon}\gamma^\mu u_p \stackrel{k\to 0}{=} \frac{2p^\mu-k^\mu}{k^2-2kp+i\varepsilon}\mathcal{M}^{(0)}(p)u_p. \qquad (1)$$

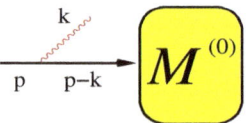

Figure 1. Kinematics of a single emission from external leg.

This way the soft-photon current $\frac{2p^\mu - k^\mu}{k^2 - 2kp + i\varepsilon}$ (either real or virtual) decouples from the hard process $\mathcal{M}$.

Let us now consider the toy-model situation shown in Figure 2. This figure corresponds to the following expression in which we have omitted all the details except for the photonic radiation in the soft approximation and the propagator of the resonance (mass M includes width Γ):

$$\mathcal{M}_N^{(0)\mu_1,\ldots,\mu_N}(k_1,\ldots,k_N) \simeq$$
$$\simeq \sum_{l=0}^{N} \sum_{\pi}^{N!} \left(\frac{2p_a^{\mu_1}}{2p_a k_{\pi_1}} \frac{2p_a^{\mu_2}}{2p_a k_{\pi_1} + 2p_a k_{\pi_2}} \cdots \frac{2p_a^{\mu_l}}{2p_a k_{\pi_1} + 2p_a k_{\pi_2} + \cdots + 2p_a k_{\pi_l}} \right)$$
$$\times \left(\frac{-2p_c^{\mu_{l+1}}}{2p_c k_{\pi_{l+1}}} \frac{-2p_c^{\mu_{l+2}}}{2p_c k_{\pi_{l+1}} + 2p_c k_{\pi_{l+2}}} \cdots \frac{-2p_c^{\mu_N}}{2p_c k_{\pi_{l+1}} + 2p_c k_{\pi_{l+2}} + \cdots + 2p_c k_{\pi_N}} \right) \quad (2)$$
$$\times \frac{1}{p_{cd}^2 - M^2}.$$

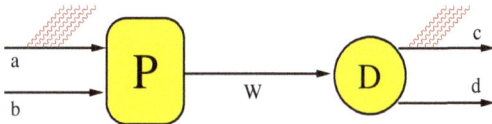

Figure 2. The soft-photon emission from external legs in the single W toy model.

The index l corresponds to the number of photons in the initial state (a) and π denotes a permutation of photons. In the first step we use the well known algebraical formula

$$\sum_{perm.} \frac{1}{pk_1(pk_1 + pk_2)\ldots(pk_1 + pk_2 + \ldots pk_n)} = \frac{1}{pk_1 pk_2 \ldots pk_n} \quad (3)$$

to get

$$\mathcal{M}_N^{(0)\mu_1,\ldots,\mu_N}(k_1,\ldots,k_N) \simeq \frac{1}{p_{cd}^2 - M^2} \sum_{l=0}^{N} \sum_{\pi/\pi_l/\pi_{N-l}}^{N!/l!/(n-l)!} \left(\prod_{i=1}^{l} \frac{2p_a^{\mu_i}}{2p_a k_{\pi_i}} \right) \left(\prod_{i=1}^{N-l} \frac{-2p_c^{\mu_{l+i}}}{2p_c k_{\pi_{l+i}}} \right). \quad (4)$$

The expression $\pi/\pi_l/\pi_{N-l}$ means that summation over separate permutations within the initial and final photon subgroups has been carried out. The leftover sum over permutations can be rearranged now into the sum over partitions $\wp_i$ in which each photon is labelled only as initial (a) or final (c):

$$\sum_{\substack{l_a,l_c=0\\l_a+l_c=n}}^{n} \sum_{\pi/\pi_a/\pi_b/\pi_c}^{n!/l_a!/l_c!} = \sum_{\wp=(a,c)^n}^{2^n}. \quad (5)$$

This brings us to the final result:

$$\mathcal{M}_N^{(0)\mu_1,\ldots,\mu_N}(k_1,\ldots,k_N) \simeq \frac{1}{p_{cd}^2 - M^2} \sum_{\wp=(a,c)^N}^{2^N} \left(\prod_{i=1}^{N} \frac{2\theta_{\wp_i} p_{\wp_i}^{\mu_i}}{2p_{\wp_i} k_i} \right). \quad (6)$$

The θ_i is equal to $+1(-1)$ for the initial (final) photon. The formula (6) can also be conveniently rewritten by turning the sum over partitions into a product, e.g., for two photons we have:

$$\frac{p_a^{\mu_1}}{p_a k_1} \frac{p_a^{\mu_2}}{p_a k_2} - \frac{p_a^{\mu_1}}{p_a k_1} \frac{p_c^{\mu_2}}{p_c k_2} - \frac{p_c^{\mu_1}}{p_c k_1} \frac{p_a^{\mu_2}}{p_a k_2} + \frac{p_c^{\mu_1}}{p_c k_1} \frac{p_c^{\mu_2}}{p_c k_2} = \left(\frac{p_a^{\mu_1}}{p_a k_1} - \frac{p_c^{\mu_1}}{p_c k_1} \right) \left(\frac{p_a^{\mu_2}}{p_a k_2} - \frac{p_c^{\mu_2}}{p_c k_2} \right). \quad (7)$$

This way we obtain the formula in the form of YFS61:

$$\mathcal{M}_N^{(0)\mu_1,\ldots,\mu_N}(k_1,\ldots,k_N) \simeq \prod_{i=1}^{N}\left(\frac{2p_a^{\mu_i}}{2p_a k_i} - \frac{2p_c^{\mu_i}}{2p_c k_i}\right). \tag{8}$$

Squaring and integrating over k_i with the $1/N!$ Bose–Einstein symmetry factor leads us to the desired exponential form. Note that, contrary to Equation (6), the formula (8) will not be valid when the recoil effect is included, see below for more comments.

2.2. Inclusion of Emission from W Boson

Now let us extend the analysis to the situation depicted in Figure 3. We include multiple emission of l_n soft photons from an internal W line. Emissions are connected by the W propagators. If we consider a product of two W propagators we find that it can be rewritten as a sum of two terms, each of them being a product of a single W propagator and the soft-photon emission factor before or after the W.

$$\frac{1}{(Q_0^2 - M^2)}\frac{1}{(Q_1^2 - M^2)} = \frac{1}{(Q_0^2 - Q_1^2)(Q_1^2 - M^2)} - \frac{1}{(Q_0^2 - M^2)(Q_0^2 - Q_1^2)}$$

$$= \frac{1}{(2k_1 Q_0 - k_1^2)(Q_1^2 - M^2)} - \frac{1}{(Q_0^2 - M^2)(2k_1 Q_1 + k_1^2)}. \tag{9}$$

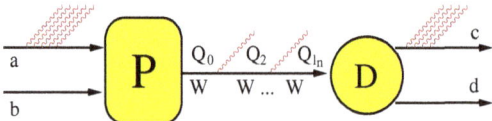

Figure 3. The soft-photon emissions from external legs and from the W propagator in the single W toy model.

That formula has a simple physical interpretation: soft emission belongs either to production or to decay phase and the W propagator knows it because its four-momentum is correctly adjusted (recoil effect). We can now generalize it to l_n emissions shown in Figure 3:

$$\sum_{permut.}\frac{1}{(Q_0^2 - M^2)(Q_1^2 - M^2)\ldots(Q_{l_n}^2 - M^2)}$$
$$= \sum_{\wp=(P,D)^{l_n}}\prod_{\wp_i=P}\frac{1}{(Q_\wp + k_i)^2 - Q_\wp^2} \times \frac{1}{Q_\wp^2 - M^2} \times \prod_{\wp_i=D}\frac{1}{(Q_\wp - k_i)^2 - Q_\wp^2}, \tag{10}$$

where $Q_\wp = Q_0 - \sum_{\wp_i=P} k_i = Q_{l_n} + \sum_{\wp_i=D} k_i$. As before, the physical picture, illustrated in Figure 4, is clear: a set of soft emissions (interpreted as belonging to the production process) precedes the W propagator and a second set (interpreted as belonging to the decay process) follows it. The recoil is properly included in the W propagator.

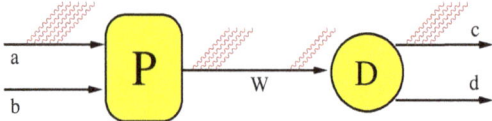

Figure 4. The soft-photon emission from external legs and W boson factorized into the production and decay multiple soft-photon emissions and one W propagator.

As for the numerators of the W propagators, in the soft limit and in the on-shell approximation (numerators are mild functions of momenta, so accounting for the recoil is not needed), we find

$$D_W(p) V(p, k, p-k)_\rho D_W(p) \stackrel{k \to 0, p^2 \to M_W^2}{=} D_W(p)(-2p_\rho), \tag{11}$$

where $V(p, k, p-k)_\rho$ is the $W\gamma W$ vertex and $D_W(p)$ denotes the numerator of the W propagator. For more emissions Equation (11) has a self-repeating structure and reduces multiple emissions into numerators of the soft factors, i.e., $\prod_i 2p_{\rho_i}$.

Combining Equation (2) with (10) and (11) we can write down the formula that corresponds to Figure 4:

$$\mathfrak{M}_N^{(0)\,\mu_1,\ldots,\mu_N}(k_1,\ldots,k_N) \simeq \sum_{\substack{l_a, l_c, n=0 \\ l_a+l_c+n=N}}^{N} \sum_{\substack{l_g, l_h=0 \\ l_g+l_h=n}}^{n} \sum_{\pi}^{N!} \tag{12}$$

$$\left[\left(\frac{2p_a^{\mu_{\pi_1}}}{2p_a k_{\pi_1}} \frac{2p_a^{\mu_{\pi_2}}}{2p_a k_{\pi_1} + 2p_a k_{\pi_2}} \cdots \frac{2p_a^{\mu_{\pi_{l_a}}}}{2p_a k_{\pi_1} + 2p_a k_{\pi_2} + \cdots + 2p_a k_{\pi_{l_a}}} \right) \right. \tag{13}$$

$$\left(\frac{-2Q_{\pi_0}^{\mu_{\pi_{l_a+1}}}}{2Q_{\pi_{l_g}} k_{\pi_{l_a+1}}} \frac{-2Q_{\pi_0}^{\mu_{\pi_{l_a+2}}}}{2Q_{\pi_{l_g}} k_{\pi_{l_a+1}} + 2Q_{\pi_{l_g}} k_{\pi_{l_a+2}}} \cdots \frac{-2Q_{\pi_0}^{\mu_{\pi_{l_a+l_g}}}}{2Q_{\pi_{l_g}} k_{\pi_{l_a+1}} + 2Q_{\pi_{l_g}} k_{\pi_{l_a+2}} + \cdots + 2Q_{\pi_{l_g}} k_{\pi_{l_a+l_g}}} \right) \tag{14}$$

$$\frac{D_W(Q_{\pi_{l_g}})}{Q_{\pi_{l_g}}^2 - M_W^2} \tag{15}$$

$$\left(\frac{2Q_{\pi_0}^{\mu_{\pi_{l_a+l_g+1}}}}{2Q_{\pi_{l_g}} k_{\pi_{l_a+l_g+1}}} \frac{2Q_{\pi_0}^{\mu_{\pi_{l_a+l_g+2}}}}{2Q_{\pi_{l_g}} k_{\pi_{l_a+l_g+1}} + 2Q_{\pi_{l_g}} k_{\pi_{l_a+l_g+2}}} \cdots \right. \tag{16}$$

$$\left. \frac{2Q_{\pi_0}^{\mu_{\pi_{l_a+l_g+l_h}}}}{2Q_{\pi_{l_g}} k_{\pi_{l_a+l_g+1}} + 2Q_{\pi_{l_g}} k_{\pi_{l_a+l_g+2}} + \cdots + 2Q_{\pi_{l_g}} k_{\pi_{l_a+l_g+l_h}}} \right) \tag{17}$$

$$\left. \left(\frac{-2p_c^{\mu_{\pi_{l_a+n+1}}}}{2p_c k_{\pi_{l_a+n+1}}} \frac{-2p_c^{\mu_{\pi_{l_a+n+2}}}}{2p_c k_{\pi_{l_a+n+1}} + 2p_c k_{\pi_{l_a+n+2}}} \cdots \frac{-2p_c^{\mu_{\pi_{l_a+n+l_c}}}}{2p_c k_{\pi_{l_a+n+1}} + 2p_c k_{\pi_{l_a+n+2}} + \cdots + 2p_c k_{\pi_{l_a+n+l_c}}} \right) \right]. \tag{18}$$

Let us explain this long formula. Lines (13) and (18) describe the standard YFS emission from the external legs a and c, whereas lines (14), (16) and (17) describe the emissions from the W-boson. It is important that both groups have identical structure and we can re-interpret them as the standard YFS emission in the production phase (lines (13) and (14)) and in the decay (lines (16) and (18)). Standard resummations can now be performed separately on the production and on the decay in complete analogy to Equation (8).

This is possible because the recoil does not depend separately on photons from electron (muon) and from the W boson, but only on their sum.

The production and decay parts are still interconnected by the sum over partitions of photons between production and decay of the form of Equation (6). This leads to the final formula

$$\mathfrak{M}_N^{(0)} \simeq \sum_{\wp=(P,D)^N}^{2^N} \frac{D_W(Q_g)}{Q_g^2 - M_W^2} \prod_{i=1}^{N} j_{\wp_i}^{\mu_i},$$

$$j_P^{\mu_i} = \frac{2p_a^{\mu_i}}{2p_a k_i} - \frac{2Q_g^{\mu_i}}{2Q_g k_i}, \quad j_D^{\mu_i} = \frac{2Q_g^{\mu_i}}{2Q_g k_i} - \frac{2p_c^{\mu_i}}{2p_c k_i}, \quad Q_g = p_{cd} + \sum_{\text{decay}} k_i. \tag{19}$$

$\sum_{\wp=(P,D)^N}^{2^N}$ is a sum over partitions of photons emitted in the production and in the decay.

3. W-Pair Production and Decay

Having explained in very detail the resummation of the real emission in the toy model, we now proceed briefly to the resummation in the W-pair production and decay process. Let us include virtual photons into the formulae. In the case of YFS61, shown on LHS of Figure 5, their resummation goes in complete analogy to the real photons, and the appropriate formula for m real photons and an arbitrary number of virtual ones, for 6 external particles, reads:

$$M^{\mu_1\mu_2\cdots\mu_m}(k_1,k_2,\ldots,k_m) =$$
$$= \mathcal{M} \prod_{l=1}^{m} j^{\mu_l}(k_l) \sum_{n=0}^{\infty} \frac{1}{n!} \prod_{i=1}^{n} \int \frac{i}{(2\pi)^3} \frac{d^4k_i}{k_i^2 - \lambda^2 + i\varepsilon} J_6^\mu(k_i) x \circ J_{6\mu}(k_i)$$
$$= \mathcal{M} \prod_{l=1}^{m} j^{\mu_l}(k_l) \, e^{\alpha B_6}, \tag{20}$$
$$B_6 = \int \frac{i}{(2\pi)^3} \frac{d^4k}{k^2 - \lambda^2 + i\varepsilon} J_6(k) \circ J_6(k),$$

where

$$J_6^\mu(k) = \sum_{X=a,b,c,d,e,f} \hat{J}_X^\mu(k),$$
$$\hat{J}_X^\mu(k) \equiv Q_X \theta_X \frac{2p_X^\mu \theta_X - k^\mu}{k^2 - 2p_X k \theta_X + i\varepsilon}, \tag{21}$$
$$J_X(k) \circ J_Y(k) = J_X(k) J_Y(-k) \text{ for } X \neq Y, \quad J_X(k) \circ J_X(k) = J_X(k) J_X(k).$$

The real soft-photon emissions have the familiar form of a product of the currents $j^{\mu_l}(k_l)$, and the similar virtual soft-photon currents $J^\mu(k)$ are resummed.

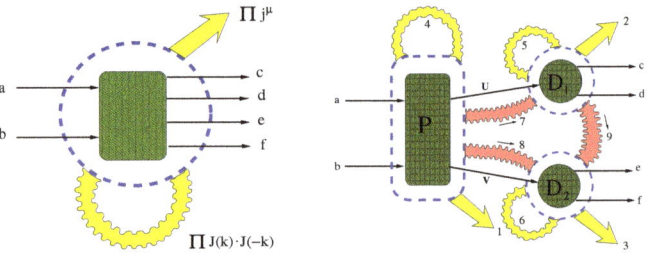

Figure 5. LHS: the classical YFS61 soft photon emission in W-pair production and decay. RHS: a new extended scheme with real and virtual emissions from Ws and all interferences, P–D and D–D.

Let us now proceed to a new, extended scheme. That scheme is illustrated on RHS of Figure 5. We have additional real soft-photon emissions from Ws as well as all virtual soft-photon interferences between the production and two decays. The resummation of real emissions proceeds exactly as in the toy example as a product of currents. The virtual soft-photon emissions follow the same logics and formulae as the real ones. The only exceptions are the issues related to the definition of the mass and width of the resonance and the UV renormalisation. Analysis of these issues is beyond the scope of this paper. Here we make an educated guess based on the solid principle of the cancellation of soft-photon singularities in QED and propose the following formula for a given partition of real momenta $k_{i_1}, k_{i_2}, k_{i_3}$

$$M_{n_1 n_2 n_3}^{\mu_{1_1} \ldots \mu_{3 n_3}}(\{k\}) = \mathcal{M}_0 \prod_{i_1=1}^{n_1} j_P^{\mu_{i_1}}(k_{i_1}) \prod_{i_2=1}^{n_2} j_{D_1}^{\mu_{i_2}}(k_{i_2}) \prod_{i_3=1}^{n_3} j_{D_2}^{\mu_{i_3}}(k_{i_3}) \times$$

$$\sum_{n_4=0}^{\infty} \frac{1}{n_4!} \prod_{i_4=1}^{n_4} \int \frac{i}{(2\pi)^3} \frac{d^4 k_{i_4}}{k_{i_4}^2 - m_\gamma^2 + i\varepsilon} J_P(k_{i_4}) \circ J_P(k_{i_4})$$

$$\sum_{n_5=0}^{\infty} \frac{1}{n_5!} \prod_{i_5=1}^{n_5} \int \frac{i}{(2\pi)^3} \frac{d^4 k_{i_5}}{k_{i_5}^2 - m_\gamma^2 + i\varepsilon} J_{D_1}(k_{i_5}) \circ J_{D_1}(k_{i_5})$$

$$\sum_{n_6=0}^{\infty} \frac{1}{n_6!} \prod_{i_6=1}^{n_6} \int \frac{i}{(2\pi)^3} \frac{d^4 k_{i_6}}{k_{i_6}^2 - m_\gamma^2 + i\varepsilon} J_{D_2}(k_{i_6}) \circ J_{D_2}(k_{i_6}) \qquad (22)$$

$$\sum_{n_7=0}^{\infty} \frac{1}{n_7!} \prod_{i_7=1}^{n_7} 2 \int \frac{i}{(2\pi)^3} \frac{d^4 k_{i_7}}{k_{i_7}^2 - m_\gamma^2 + i\varepsilon} J_P(k_{i_7}) \circ J_{D_1}(k_{i_7})$$

$$\sum_{n_8=0}^{\infty} \frac{1}{n_8!} \prod_{i_8=1}^{n_8} 2 \int \frac{i}{(2\pi)^3} \frac{d^4 k_{i_8}}{k_{i_8}^2 - m_\gamma^2 + i\varepsilon} J_P(k_{i_8}) \circ J_{D_2}(k_{i_8})$$

$$\sum_{n_9=0}^{\infty} \frac{1}{n_9!} \prod_{i_9=1}^{n_9} 2 \int \frac{i}{(2\pi)^3} \frac{d^4 k_{i_9}}{k_{i_9}^2 - m_\gamma^2 + i\varepsilon} J_{D_1}(k_{i_9}) \circ J_{D_2}(k_{i_9})$$

$$\frac{1}{(p_{cd} + K_2 - K_7 + K_9)^2 - M^2} \frac{1}{(p_{ef} + K_3 - K_8 - K_9) - M^2}.$$

$K_m = \sum_{i_m} k_{i_m}$ is the sum of appropriate photon momenta as depicted on the RHS of Figure 5. The recoiled W propagators do not depend on K_4, K_5 and K_6, i.e. on the interferences $P - P, D_1 - D_1$ and $D_2 - D_2$, so the corresponding sums can be folded into the exponential form. For example:

$$\sum_{n_4=0}^{\infty} \frac{1}{n_4!} \prod_{i_4=1}^{n_4} \alpha B_P = e^{\alpha B_P}, \quad \alpha B_P = \int \frac{i}{(2\pi)^3} \frac{d^4 k_{i_4}}{k_{i_4}^2 - m_\gamma^2 + i\varepsilon} J_P(k_{i_4}) \circ J_P(k_{i_4}). \qquad (23)$$

In order to fold the virtual sums 7, 8 and 9 we have to rearrange the W propagators which depend on K_7, K_8 and K_9. That can be done in the soft-photon approximation, i.e., dropping all bilinear products of the type $k_i k_j$. With the help of formulae such as

$$\left(1 + \sum \kappa_i\right)^2 \simeq \prod (1 + \kappa_i)^2 + \mathcal{O}(\kappa_i \kappa_j) \qquad (24)$$

we can write for one of the Ws

$$\frac{U_2^2 - M^2}{(U_2 - K_7 + K_9)^2 - M^2} \simeq \prod_{i_7} \frac{U_2^2 - M^2}{(U_2 - k_{i_7})^2 - M^2} \prod_{i_9} \frac{U_2^2 - M^2}{(U_2 + k_{i_9})^2 - M^2}, \qquad (25)$$

where $U_2 = p_{cd} + K_2 = p_{cd} + \sum_{D_1} k_{i_2}$. That way we have rewritten the recoiled W propagator in a factorized form suitable for the resummation and we can write down the final formula

$$M^{\mu_1 \ldots \mu_n}(k_1, k_2, \ldots, k_n) = \sum_{\wp \in (P, D_1, D_2)^n} \mathcal{M}_0 \prod_{i=1}^{n} j_{\wp_i}^{\mu_i}(k_i) e^{\alpha B_{10}^{\text{CEEX}}(U_\wp, V_\wp)} \frac{1}{U_\wp^2 - M^2} \frac{1}{V_\wp^2 - M^2}, \qquad (26)$$

$$U_\wp = p_{cd} + \sum_{\wp_i = D_1} k_i, \quad V_\wp = p_{ef} + \sum_{\wp_i = D_2} k_i.$$

The B_{10} function is defined as

$$\begin{aligned}
\alpha B_{10}^{\mathrm{CEEX}}(U,V) &= \alpha B_P + \alpha B_{D_1} + \alpha B_{D_2} \\
&\quad + 2\alpha B_{P\otimes D_1}(U) + 2\alpha B_{P\otimes D_2}(V) + 2\alpha B_{D_1\otimes D_2}(U,V), \\
\alpha B_{P\otimes D_1}(U) &= \int \frac{i}{(2\pi)^3} \frac{d^4k}{k^2 - m_\gamma^2 + i\varepsilon}\, J_P(k) \circ J_{D_1}(k) \frac{U^2 - M^2}{(U-k)^2 - M^2}, \\
\alpha B_{P\otimes D_2}(V) &= \int \frac{i}{(2\pi)^3} \frac{d^4k}{k^2 - m_\gamma^2 + i\varepsilon}\, J_P(k) \circ J_{D_2}(k) \frac{V^2 - M^2}{(V-k)^2 - M^2}, \\
\alpha B_{D_1\otimes D_2}(U,V) &= \int \frac{i}{(2\pi)^3} \frac{d^4k}{k^2 - m_\gamma^2 + i\varepsilon}\, J_{D_1}(k) \circ J_{D_2}(k) \frac{U^2 - M^2}{(U+k)^2 - M^2} \frac{V^2 - M^2}{(V-k)^2 - M^2}.
\end{aligned} \quad (27)$$

Equations (26) and (27) are the principal new result presented in this note. Equation (26) has an identical, exponential form as the original YFS61 formula of Equation (20). The only difference is that the B functions responsible for the virtual interferences include now ratios of W propagators, see Equation (27). That way the recoil effect is incorporated into the scheme.

4. Summary

In this note we reviewed the soft photon exponentiation of YFS61 and proposed its extension to the case with charged semi-stable internal resonances. We explained in a combinatorial way how the YFS61 resummation of real radiation proceeds on an example of a toy model $e\bar{\nu} \to W \to \mu\bar{\nu}$ and then we introduced real radiation from the W-boson. Virtual emissions we included in a form of educated guess based on a firm ground of cancellations of soft-photon singularities in QED. This has been done for the full scale, FCCee motivated, process $ee \to WW \to 4fermions$. The result of the analysis is a formula which generalizes the YFS61 scheme. In its form it is identical to the original one, i.e., has the exponential form for virtual emissions and the sum over partitions for the real emissions. The difference is in the shape of the virtual B functions which acquire dependence on the W propagators. The formula has two basic features: it is exact in the soft-photon limit and it includes recoil of the W propagators. The proposed exponentiation can be a solid starting point for a construction of a new generation of Monte Carlo algorithms for the increased precision needed by the FCCee project.

Author Contributions: All authors contributed equally to the reported research.

Funding: This work is partly supported by the Polish National Science Center grant 2016/23/B/ST2/03927 and the CERN FCC Design Study Programme.

Conflicts of Interest: The authors declare no conflict of interest.

References

1. Yennie, D.R.; Frautschi, S.C.; Suura, H. The infrared divergence phenomena and high-energy processes. *Ann. Phys.* **1961**, *13*, 379–452. [CrossRef]
2. Greco, M.; Pancheri-Srivastava, G.; Srivastava, Y. Radiative Corrections for Colliding Beam Resonances. *Nucl. Phys.* **1975**, *B101*, 234–262. [CrossRef]
3. Greco, M.; Pancheri-Srivastava, G.; Srivastava, Y. Radiative Corrections to $e^+e^- \to \mu^+\mu^-$ Around the Z0. *Nucl. Phys.* **1980**, *B171*, 118. [CrossRef]
4. Jadach, S.; Ward, B.F.L.; Was, Z. Coherent exclusive exponentiation for precision Monte Carlo calculations. *Phys. Rev.* **2001**, *D63*, 113009. [CrossRef]
5. Jadach, S.; Płaczek, W.; Skrzypek, M. QED Exponentiation for quasi-stable charged particles: The $e^-e^+ \to W^-W^+$ process. *arXiv* **2019**, arXiv:1906.09071.
6. Jadach, S. Yennie-Frautschi-Suura Soft Photons in Monte Carlo Event Generators. Unpublished, 1987. Available online: http://inspirehep.net/record/244950 (access on 15 October 2019)
7. Jadach, S.; Richter-Was, E.; Ward, B.F.L.; Was, Z. Monte Carlo program BHLUMI-2.01 for Bhabha scattering at low angles with Yennie-Frautschi-Suura exponentiation. *Comput. Phys. Commun.* **1992**, *70*, 305–344. [CrossRef]

8. Jadach, S.; Ward, B.F.L.; Was, Z. The Monte Carlo program KORALZ, version 3.8, for the lepton or quark pair production at LEP/SLC energies. *Comput. Phys. Commun.* **1991**, *66*, 276–292. [CrossRef]
9. Jadach, S.; Ward, B.F.L.; Was, Z. The Precision Monte Carlo event generator KK for two fermion final states in e^+e^- collisions. *Comput. Phys. Commun.* **2000**, *130*, 260–325. [CrossRef]
10. Jadach, S.; Płaczek, W.; Skrzypek, M.; Ward, B.F.L.; Wąs, Z. The Monte Carlo event generator YFSWW3 version 1.16 for W pair production and decay at LEP2/LC energies. *Comput. Phys. Commun.* **2001**, *140*, 432–474. [CrossRef]
11. Placzek, W.; Jadach, S. Multiphoton radiation in leptonic W-boson decays. *Eur. Phys. J.* **2003**, *C29*, 325–339. [CrossRef]

© 2019 by the authors. Licensee MDPI, Basel, Switzerland. This article is an open access article distributed under the terms and conditions of the Creative Commons Attribution (CC BY) license (http://creativecommons.org/licenses/by/4.0/).

MDPI
St. Alban-Anlage 66
4052 Basel
Switzerland
Tel. +41 61 683 77 34
Fax +41 61 302 89 18
www.mdpi.com

Symmetry Editorial Office
E-mail: symmetry@mdpi.com
www.mdpi.com/journal/symmetry